The Barai

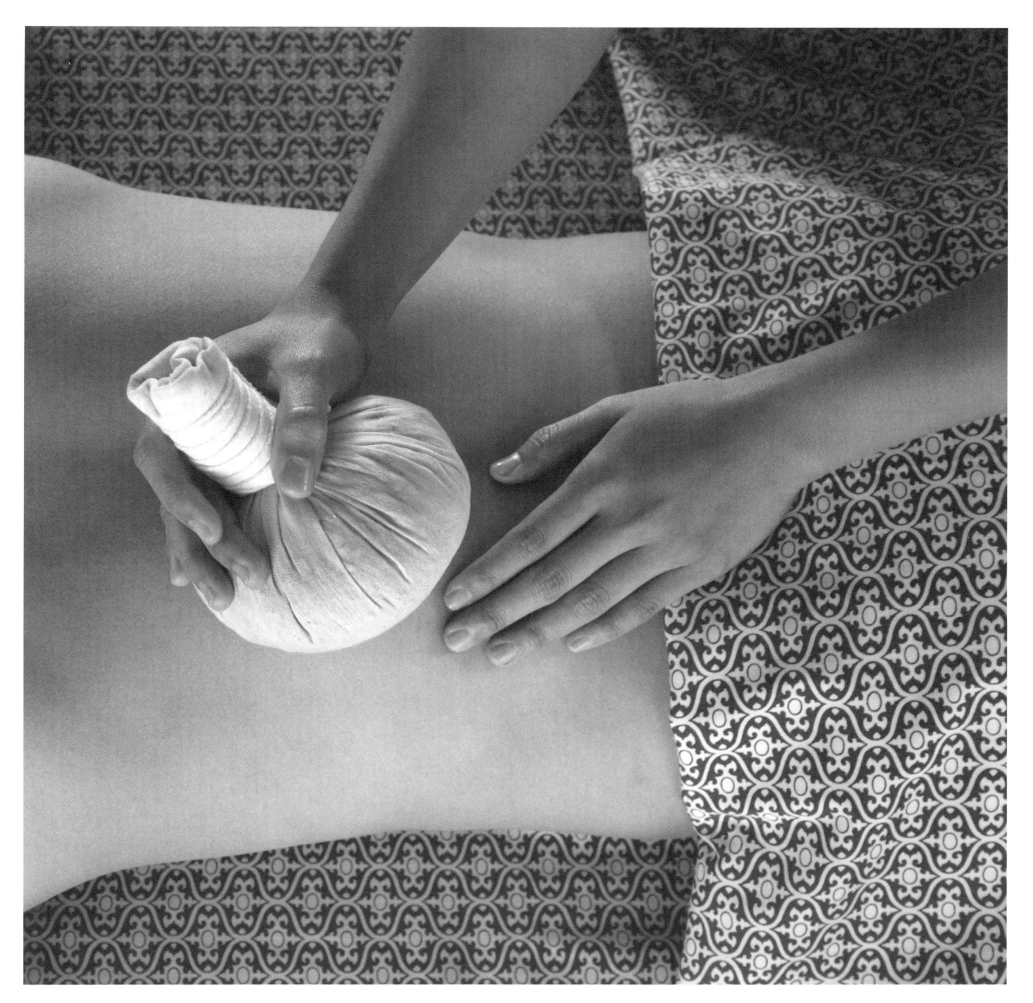

The Spa at Four Season Resort

Coqoon Spa

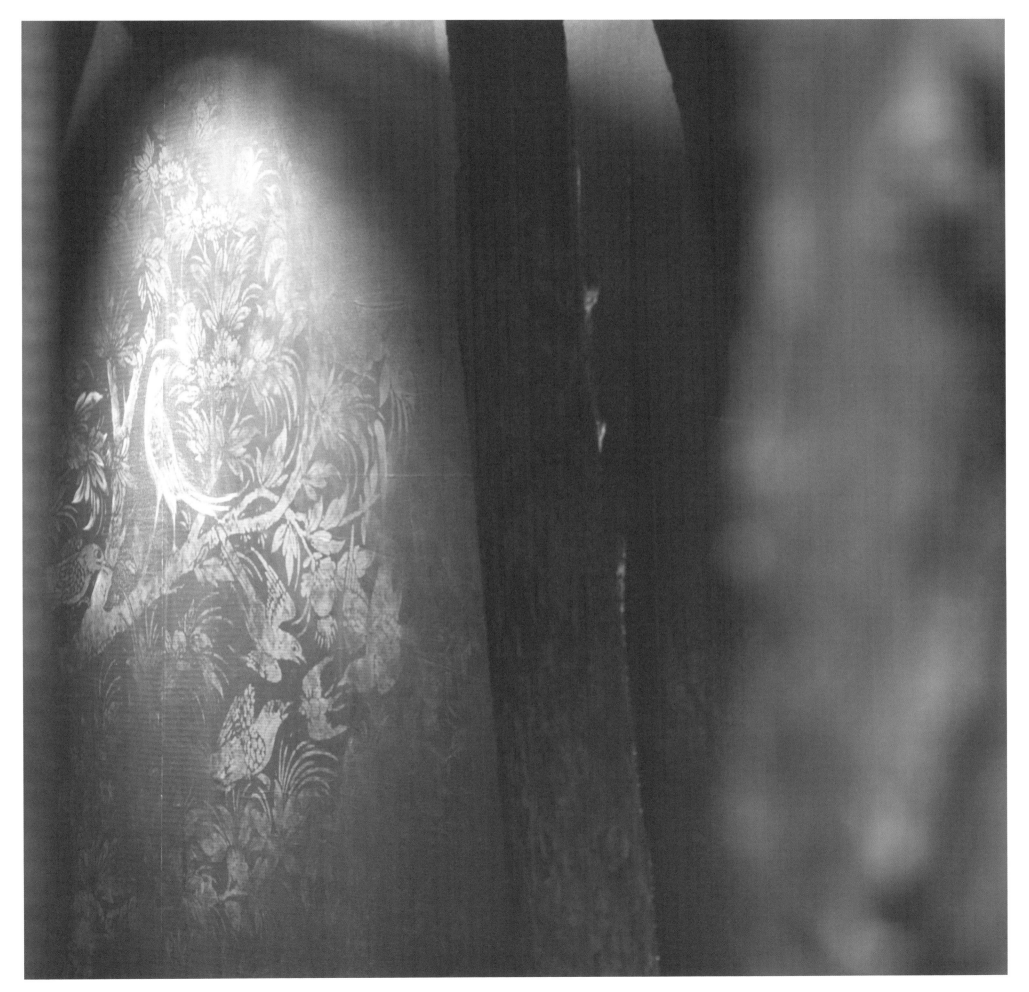

So Spa

Top Exotic Spa

in Thailand

Dalian University of Technology Press

TOP EXOTIC SPA IN THAILIAND

Copyright ©2013 by
Dalian University of Technology Press

Published by
Dalian University of Technology Press

Address: Section B, Sic-Tech Building, No.80 Software
Park Road, Ganjingzi District, Dalian, China
Tel: +86 411 84709043
Fax: +86 411 84709246
E-mail: designbooks_dutp@yahoo.cn
URL: http://www.dutp.cn

ISBN 978-7-5611-7753-2

Top Exotic Spa

in Thailand

Spa Design Development

When I was asked to write this introduction, I was pleasantly surprised to learn that at long last the spa world has started to be taken more seriously in at least one of the fields of excellence which has long been acknowledged from within the industry, but that which has yet to be rarely acknowledged by the architectural world upon whom we are so reliant to deliver on the vision of what we offer.

Spas around the world have continued to grow over recent times, with many countries reporting double digit growth rates, despite difficulties one after the other. While America is reportedly home to over a quarter of the worlds spas, the Asia Pacific region accounts for approximately one third, with Thailand being number one for the number of spas, and enjoys the highest industry value in the region, a significant statistic given that it was only in 1992 that the first spas came to the kingdom.

Since then, in leaps and bounds the Thai spa industry has grown from one which was targeted to the tourist industry, offering a luxury pampering experience, to one which is more accessible than ever, appealing not only to a massage savvy Thai clientele, but also to the seeming insatiable international consumer. The luxury pampering has also evolved, and while for many this remains an important part of the experience, the world of wellness, preventative and sustainable healthcare has become part of what is being offered, not only in Destination spas, but increasingly in the competitive hotel and resort sectors of the industry. The international tourist remains the stalwart of the Thai spa industry, an industry that probably would not exist without their support, yet both government and private enterprise continue to seek new offerings to entice spa savvy consumers in an ever more competitive world, where a visit to a spa is more about healthy living than pampering indulgence.

While many may not agree, spas around the world have a familiar offering of treatments, and while there are some which offer local or regional variations, the basics are much the same with scrubs, wraps, baths, massages and facials being the mainstay of the offering. Many have ventured to expand on this, creating customized treatments combining various therapies but there is a limit to the number of variations you can get by combining Swedish, Aromatherapy, Thai, Shiatsu, Sport massages which are the treatments most commonly taken in a spa. Beyond this it is the local culture traditions which the spa industry has raped and pillaged to its own benefit (generally without respect or consideration to those that have practiced the traditional therapies for hundreds if not thousands of years), modifying and changing the therapies so that they are practiced outside of the context from which they would traditionally be practiced, by therapists who have little or no understanding of the basis of the treatment, thereby changing history, and minimizing the potential healing benefits that could be available. Fortunately as an industry that is ever evolving, our disrespect for our forefathers is now being understood and acknowledged, with things starting to change – all be it very slowly.

Individual countries are endeavoring to define their own unique spa cultures, with many seeing Thailand as one of the most successful, give its more than 2,500 year heritage of Thai Massage, handed done from generation to generation. Today as a branch of Thai traditional medicine, able to treat a wide variety of ailments, it is regulated by the government. However the vast majority of people will never need nor experience the full scope of Thai massage techniques, besides which it is only a small proportion of practitioners whom have this level of diagnostic skill.

Today there are hundreds of schools and training centres around the kingdom and a growing number across the globe, that teach Thai massage, with all but a few teaching 'folk style' massage, well suited to the easing of minor aches and pains and the promotion of a general state of well-being. With its roots in traditional medicine, the Thai spa industry has blossomed into a unique industry, supported by culture and traditions, ancient Thai herbs and the element of Thainess into an industry that is enjoyed, copied, envied and desired, the world over.

Many in the western world have long looked to the east for inspiration, being in awe of the mystique that the region offers, so different, yet so in demand. Thailand has capitalized on this demand with its burgeoning tourism industry, developing a spa industry driven by the hotel and resort spa sector. In a market that is ever more competitive, it is the element of design in which boundaries are being pushed, as iconic spas featuring luxurious, impressive (and some may say over-indulgent) designs seem to perpetually come to the fore. Some of these spas, across the length and breadth of the kingdom, have already become iconic in their own right, be it for their design, their services, or a combination thereof, however it is the design that we are celebrating here following, to showcase that Thai spas are much, much more than Thai massage, and that design is an integral part of the overall spa experience.

The relatively new comers such as So Spa, Elemis Spa and Kempinski The Spa in the heart of Bangkok, to those that are the more established like the multi award winning Coqoon Spa and Anantara Spa (both) in Phuket; The Barai, Hua Hin and The Forest Spa, Koh Samui, they all celebrate the diversity of location, concept and design. While not always 'Thai' they are by their nature unquestionably underwritten and intrinsically liked by the Thai culture and traditions from which the industry has grown, and for which they owe much of their success. From the sense of space, emboldened by high ceilings, to muted colour schemes, and the careful selection of building materials that enhance the sense of nature (real or perceived) and promote a sense of 'whole-ness' enabling one to experience the spa on a level far surpassing that of those that seek to emulate them, that gives them the edge in a saturated market. Ultimately though, while it may give them a marketing edge it is the deliverance of quality level of service that will decide their ultimate fate in the annuls of Thai spa history.

So when visiting your next spa, take a little extra time to step back and truly look at the environment that has been created, the architectural nuances - simple or complex, natural or industrial, contemporary or traditional, or any combination thereof; and take a moment to reflect on the thought process that has been undertaken to enable the owners/developers dream to become your reality, proving you with a genuine experience as can be offered by each of the spas include in the pages following.

Andrew Jacker
President, Thai Spa Association

Top Exotic Spa in Thailand

AWAY SPA 013
W Retreat Koh Samui

ELEMIS SPA 027
The St. Regis Bagkok

EFOREA SPA AT HILTON 039
Hilton Pattaya

SO SPA 051
Sofitel So Bangkok

KEMPINSKI THE SPA 065
Siam Kempinski Hotel Bangkok

DEVARANA SPA 077
Dusit Thani Bangkok

THANN SANCTUARY 089
Crowne Plaza

THE SPA 101
The Chedi Chiang Mai

ANANTARA SPA 113
Anantara Phuket Villas

125 THE SPA
Four Seasons Resort

139 THE DHEVA SPA
Mandarin Oriental Dhara Dhevi

155 ESPA
Phulay Bay, A Ritz-Carlton Reserve

171 THE BARAI
Hyatt Regency Hua Hin

187 SALA SPA
Sala Phuket Resort and Spa

201 SPA NAKA
The Naka Island, A Luxury Collection Resort & Spa

215 COQOON SPA
Indigo Pearl Phuket

229 THE FOREST SPA
Tamarind Springs

241 PROJECT DATA

W Retreat is about escape into a world of
relaxation intermixed with a world of WOW
where understated cool meets warmth of space.

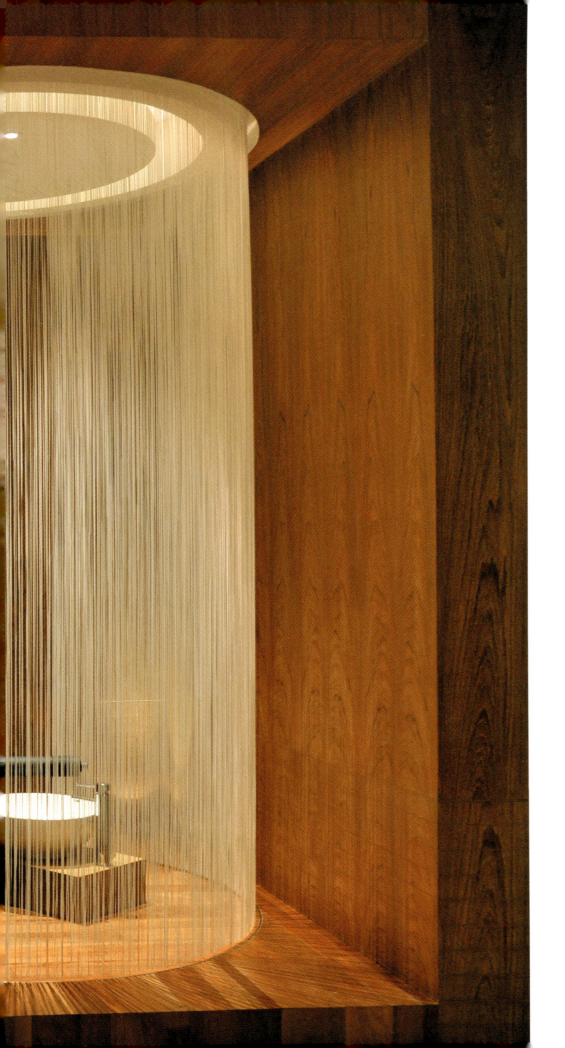

AWAY SPA

W Retreat Koh Samui, Surat Thani

Spa complex are fully decorated by natural teak wood, symbolizing Thai-ness.

Elemis Spa at the St. Regis Bangkok embraces
a warm and cosy white cocoon that feels a world
away from the bustling city below. Unwind in
a sophisticated, soothing environment of
natural light, with an exceptional view over
the Royal Sports Club.

ELEMIS SPA

The St. Regis Bangkok, Bangkok

15 luxurious treatment rooms, as well as an innovative Relaxation Zone of rain showers, Thai herbal steam rooms, vitality and foot ritual pools, sensory experience showers and relaxation nests.

eforea, the global spa experience from Hilton, combines three distinct ranges of treatments focused on organic, natural and scientific, results-driven practices, giving every type of traveler the unique therapeutic journey they seek.

EFOREA SPA AT HILTON

Hilton Pattaya, Pattaya

Drift off in modern treatment rooms to a backdrop of signature aromas and soothing music. eforea uses an array of exclusive products which harness the peaceful and restorative power of nature.

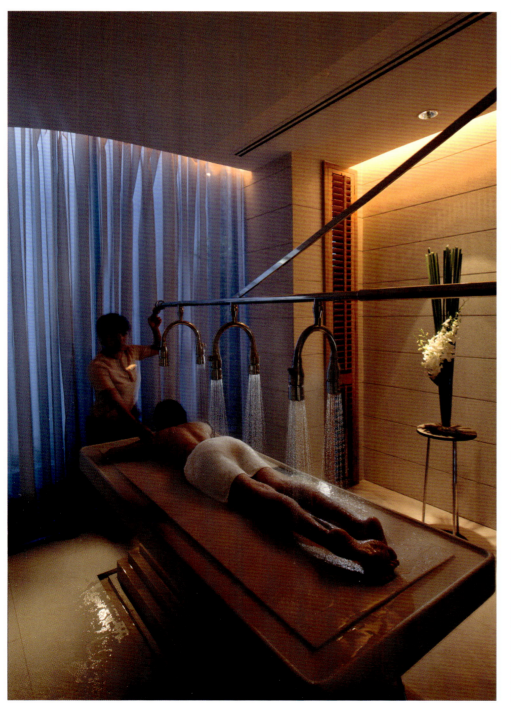

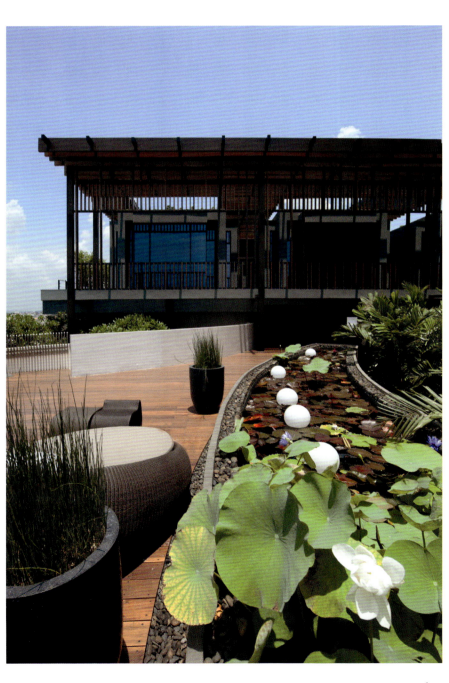

So Spa, Inspired by the Asian folktale of Himmapan. Each of the seven treatment rooms is steeped in mystique and linked to the Seven Heavenly Ponds. Retreat into the sylvan glades of bliss, with 600 square metres to uncover

SO SPA

Sofitel So Bangkok, Bangkok

So Spa's three exclusive So Suite Spa rooms are inspired by the Wood Element, with dedicated spa facilities perfect for overnight indulgence.

Tucked away like a hidden gem amid the vibrant Siam area, Kempinski The Spa is a luxurious space devoted to relaxation and the art of rejuvenation. The spa radiates a sense of tranquility that eases the stress and anxiety of guests as soon as they arrive.

KEMPINSKI THE SPA

Siam Kempinski Hotel Bangkok, Bangkok

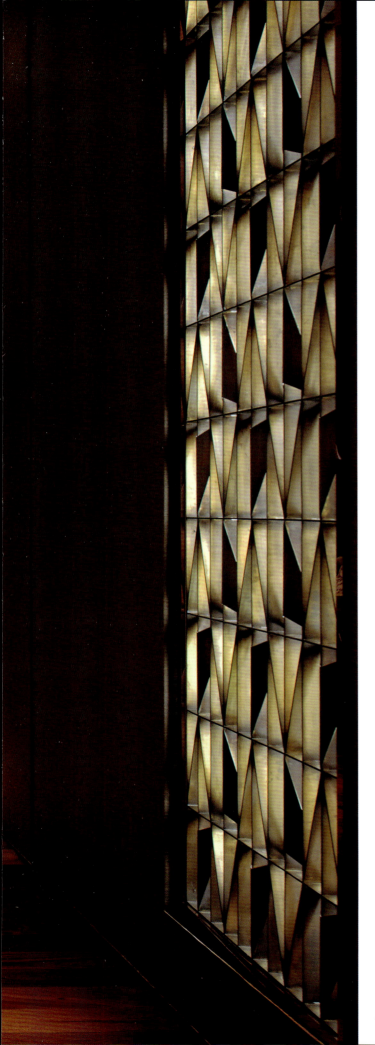

The relaxation room offers a post-treatment lounging area – a comfortable place in which to reflect on the treatment experience. The visual focus is on the artwork, with two round sculptures placed in front of a textured wall illuminated so as to induce a sense of calm and ease.

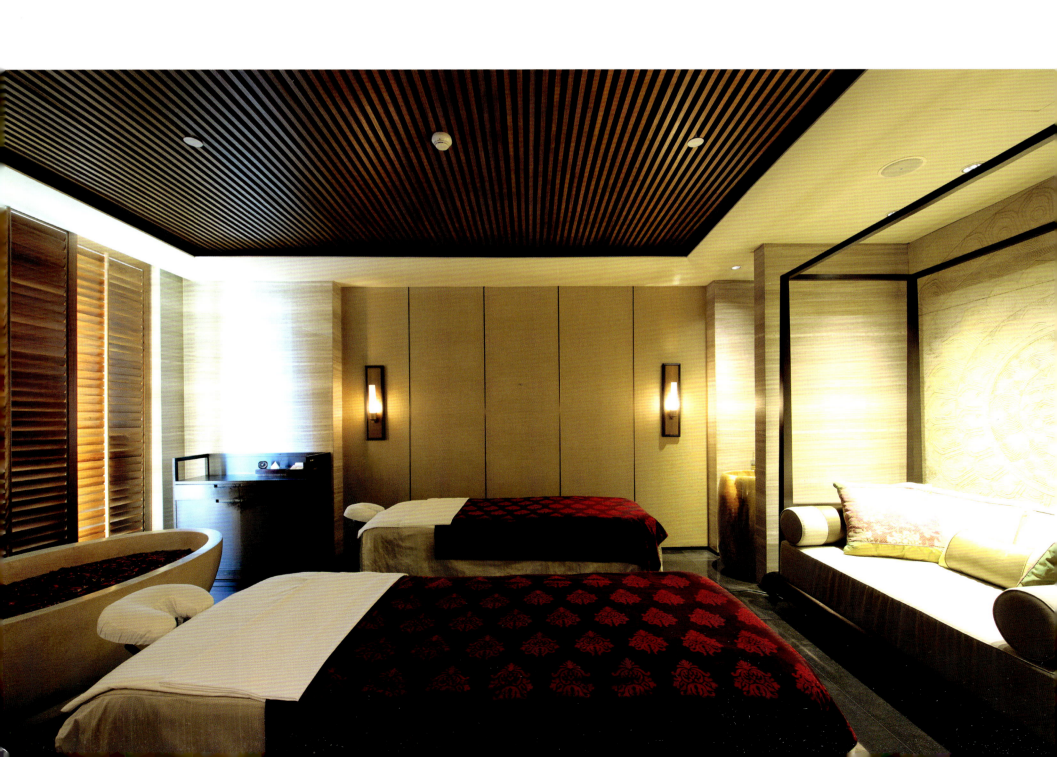

ROSEMARY

Devarana, meaning 'garden in heaven' in
Thai Sanskit, Devarana spa takes its inspiration
from an ancient Thai literary work entitled
'Tri-bhum-phra-ruang', which describes
a remarkably beautiful garden at heaven's gate,
surrounded by gardens and ponds.

DEVARANA SPA

Dusit Thani Bangkok

Devarana Spa follows the timeless concept, opening doors into a whole new world of tranquility.

Devarana's sumptuous treatment rooms take you to joyous inner sanctum.
Each suite is so gently soothed by shaded natural light and warm color scheme
that's sheer heaven as you ease onto the comfortable bed.

Thann Sanctuary, All senses will be awakened
by the unique holistic spa concept
and ambience.

THANN SANCTUARY

Crowne Plaza, Phuket

The unique and outstanding in the decoration oozes fine design and
quality to create a perfect place for physical, mental and spiritual renewal

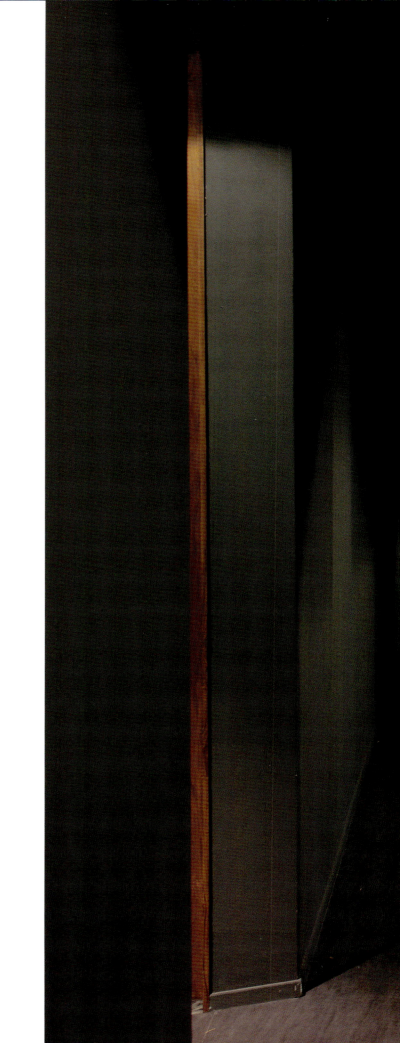

All senses will be awakened by the unique holistic spa concept and ambience.

Distinctive teak doors lead from The Spa at
The Chedi Chiang Mai's strikingly minimalist
exterior to an inner sanctum where the emphasis
is on pure holistic pampering, guest may opt
to linger on the rooftop deck overlooking the
Mae Ping or join sunbathers below on cosy
loungers that line the 34-metre swimming pool.

THE SPA

The Chedi Chiang Mai, Chiang Mai

The design intention of the scheme was to combine a city location with the atmosphere of a resort hotel, together with the integration of the existing colonial house, site and hotel using a contemporary interpretation of traditional Thai materials.

Anantara Spa Phuket offers a haven of calm
and serenity where you can leave behind
the stresses of everyday life and begin to
synchronise the subtle energies of your being.

ANANTARA SPA

Anantara Phuket Villas, Phuket

A luxurious temple of harmony for body, mind and spirit – an indulgent sanctuary created exclusively for your pleasure and well-being. The fragrance of specially blended oils, herbs and spices fills the air, and your body and mind immediately relax in anticipation of the exotic experience that awaits.

THE SPA

Four Seasons Resort, Chiang Mai

The Spa was inspired by a northern Thai temple. Exquisite interiors in rich colours of maroon, gold leaf, black and white, also feature specially commissioned artworks and sculptures.

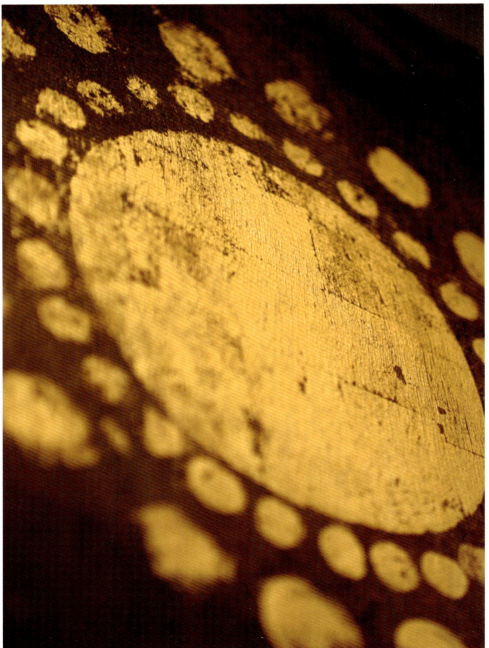

Notwithstanding its ambitious size and
the grand scale of its architecture, Mandarin
Oriental Dhara Dhevi was born out of
a very simple wish: to preserve the beauty of
northern Thailand's Lanna culture and
to create an atmosphere that would renew
and nourish the traditions, craftsmanship and
intangible charms of its people.

THE DHEVA SPA

Mandarin Oriental Dhara Dhevi, Chiang Mai

Based on the design of an ancient Mandalay palace,
the breathtaking golden teakwood architecture features a stunning
seven-tiered roof designed to represent the seven steps to
nirvana and the attainment of spiritual and physical perfection.

Every inch of immense structure is embellished with ornate moldings and sculptures depicting scared animals or symbolic Buddhist motifs, loyally recreated by 150 Chiang Mai artisans from the original Burmese template in Mandalay, Myanmar.

The spectacular exterior is just as equally matched by the sumptuous interiors where guests can enjoy impeccable service and a range of treatments drawn from three continents, with their origins spanning over 4000 years. At the same time, The Dheva Spa also draws on time-honored rites that are very much part of ancient Lanna Kingdom, of which Chiang Mai was the royal capital.

Watsu is a Healing Water Ritual – Aquatic Bodywork or 'watsu' combines oriental massage techniques and sound under water to balance one's physical and emotional tension.

Situated on the third floor of the Dheva Spa Residence, décor exudes an air of colonial elegance with five-metre high ceilings, polished teakwood floors covered with Persian carpets and beautiful furnishings including a grand piano.

Amongst the impressive backdrop of tropical jungle and the tranquility of the lime stone hills lies a peaceful haven in ESPA at Phulay Bay. Relax next to the lagoon while you await your journey into the realm of traditionally inspired Thai, Asian and European treatment experience by ESPA. Immerse yourself in the tranquility of Phulay Bay to soothe, relax and restore.

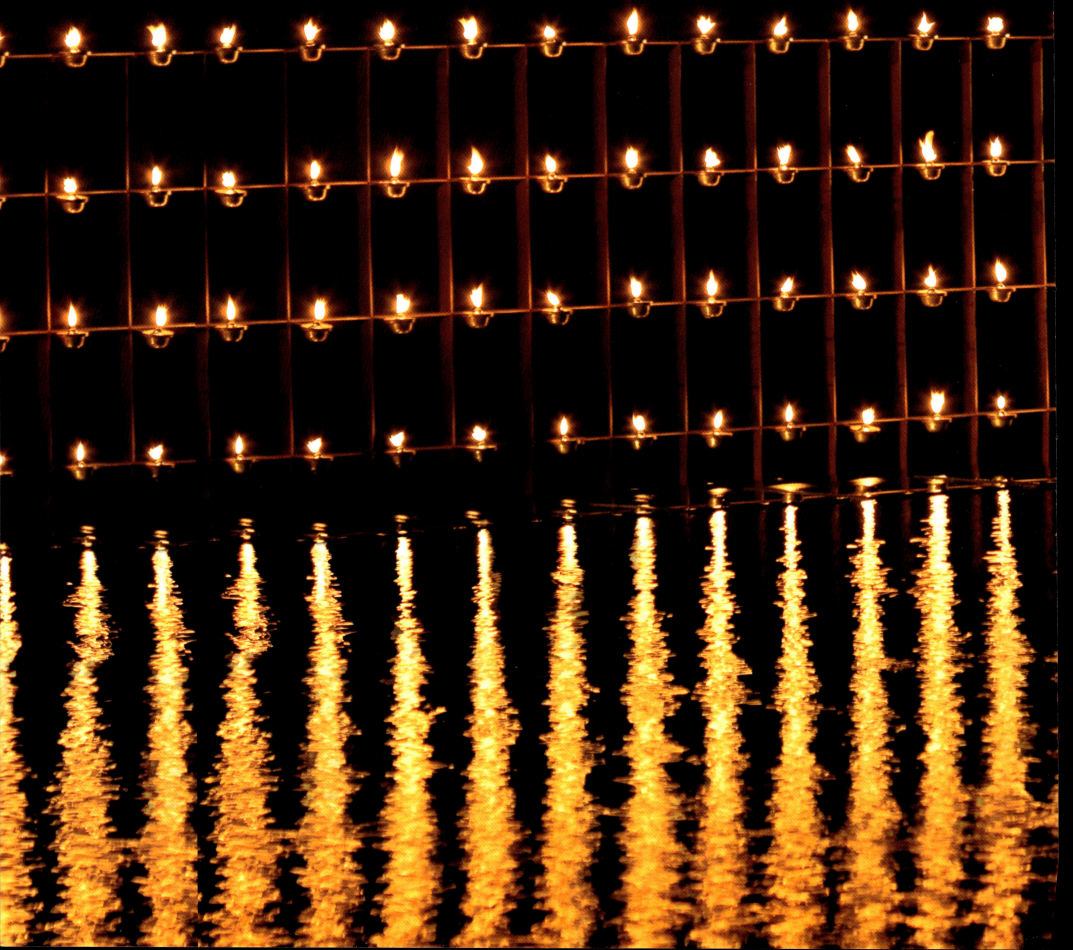

ESPA

Phulay Bay, A Ritz-Carlton Reserve, Krabi

The Spa Tea Lounge has comfortable seating for guests to wait for a treatment. Guests can also have their consultation in this area and personally escorted by a Spa Attendant to their respective male or female heat experience and relaxation room in preparation for treatments scheduled.

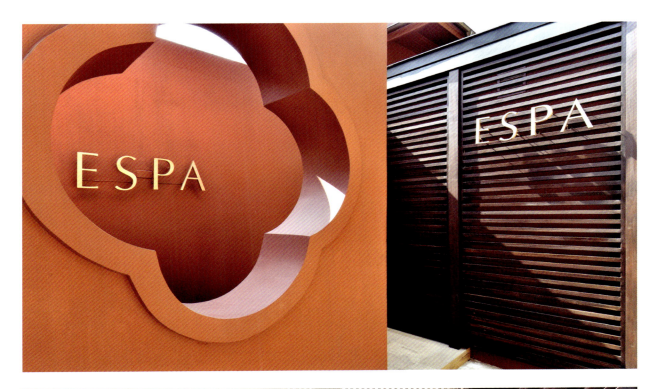

Tranquility Relaxation Areas – Offering a selection of flavored water and fresh fruits. Separate male and female lounges including 5 relaxation couches with luxurious blankets and reading material.

THE BARAI, an award-winning destination spa featuring eight spa suites and 18 treatment rooms, is inspired by rich traditions of this region. THE BARAI is a journey where the art of architecture is the guest's guide. When guests arrive at THE BARAI, they leave one world behind and enter another, where the goal of the design is to shelter guests from the outside world, offering a true sanctuary of peace and pure serenity.

THE BARAI

Hyatt Regency Hua Hin, Prachuap Khiri Khan

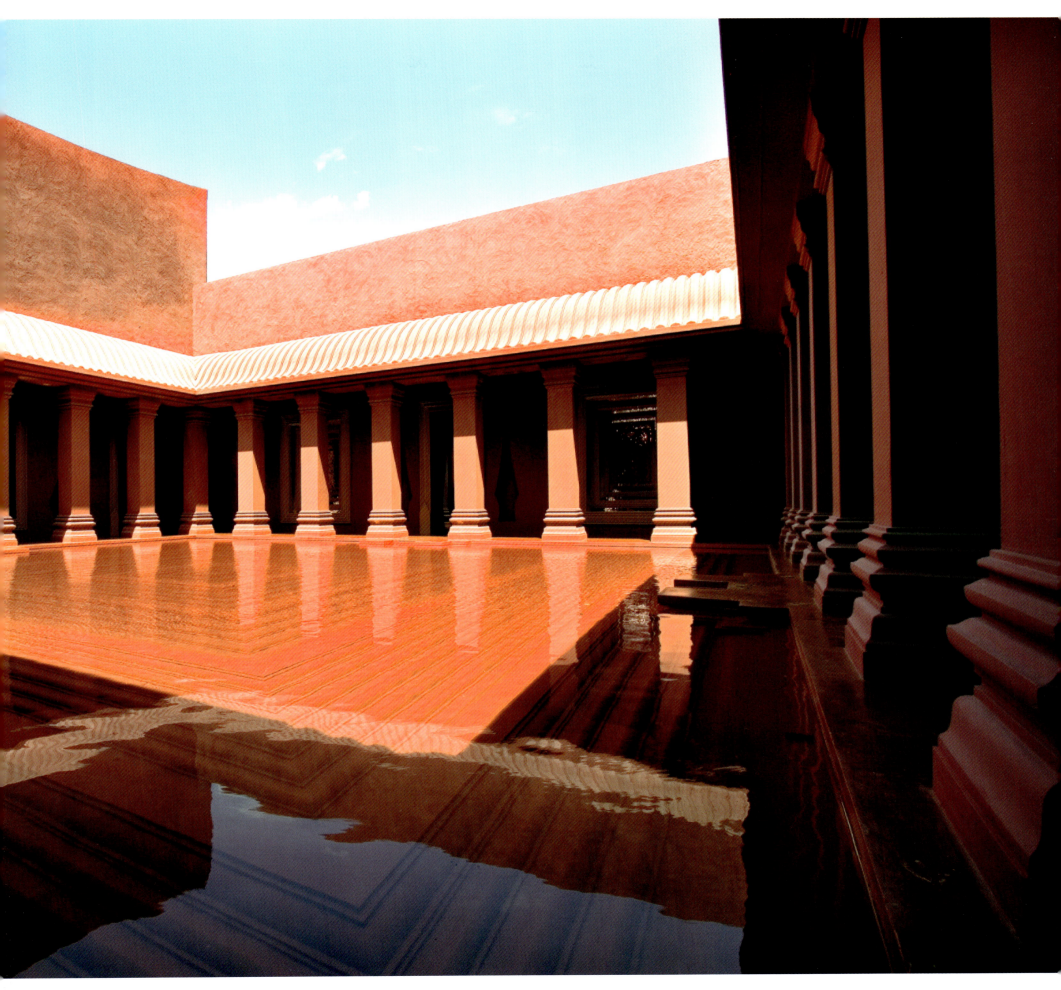

The entrance gallery to THE BARAI, dark and shadowy, with occasional playful beams of sunlight, at times narrows down and squeezes tight. Whatever their choice, they eventually find themselves under a canopy of sky in the spectacular Salarai water court, where they are welcomed and prepared for the continuation of their journey.

As a result of guests' discovery while on THE BARAI journey, THE BARAI uses architecture and interior design as a visual guide, affecting guests' feelings as they realise the goal of the journey, which is a delightful feeling of peace and tranquillity deep within the inner self.

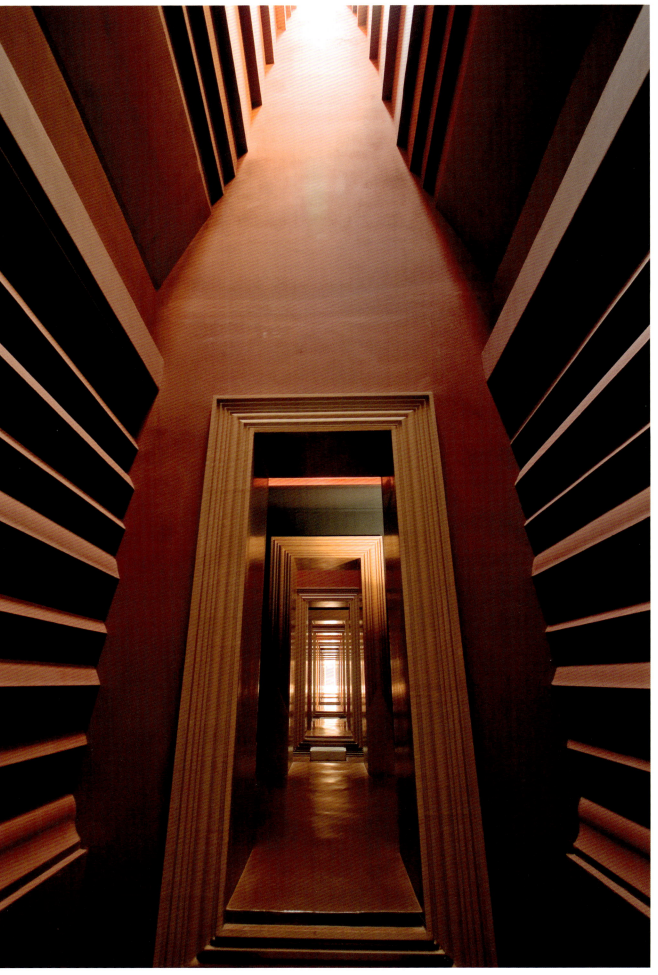

SALA Spa Phuket features contemporary Sino Portuguese inspired design evoking the sensual harmony of a tropical Chinese garden. Step into our sanctuary and be surrounded with the calming effects of water and relaxation

SALA SPA

Sala Phuket Resort and Spa, Phuket

An open-air passage approaching the spa entrance with textured cast-in-situ concrete wall details.

Spa lobby opens into an interior courtyard.

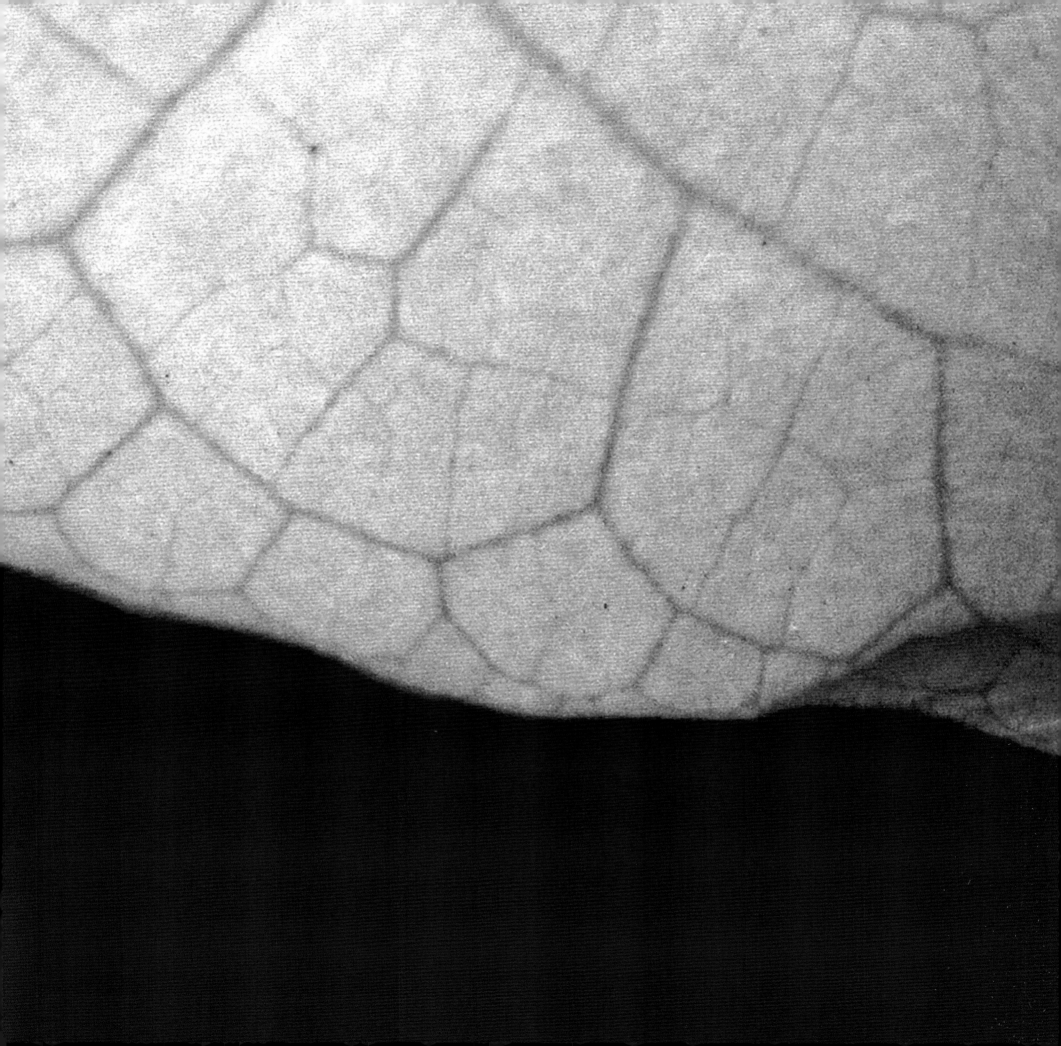

Welcome to a peaceful, tranquil sanctuary. Renew your strength and inner sense of wellbeing and experience your own private paradise and disconnect from the world.

SPA NAKA

The Naka Island, A Luxury Collection Resort & Spa, Phuket

Watsu unique experience, combining shiatsu pressure points, gentle stretches and the absence of gravity to help alleviate those aches and pains.

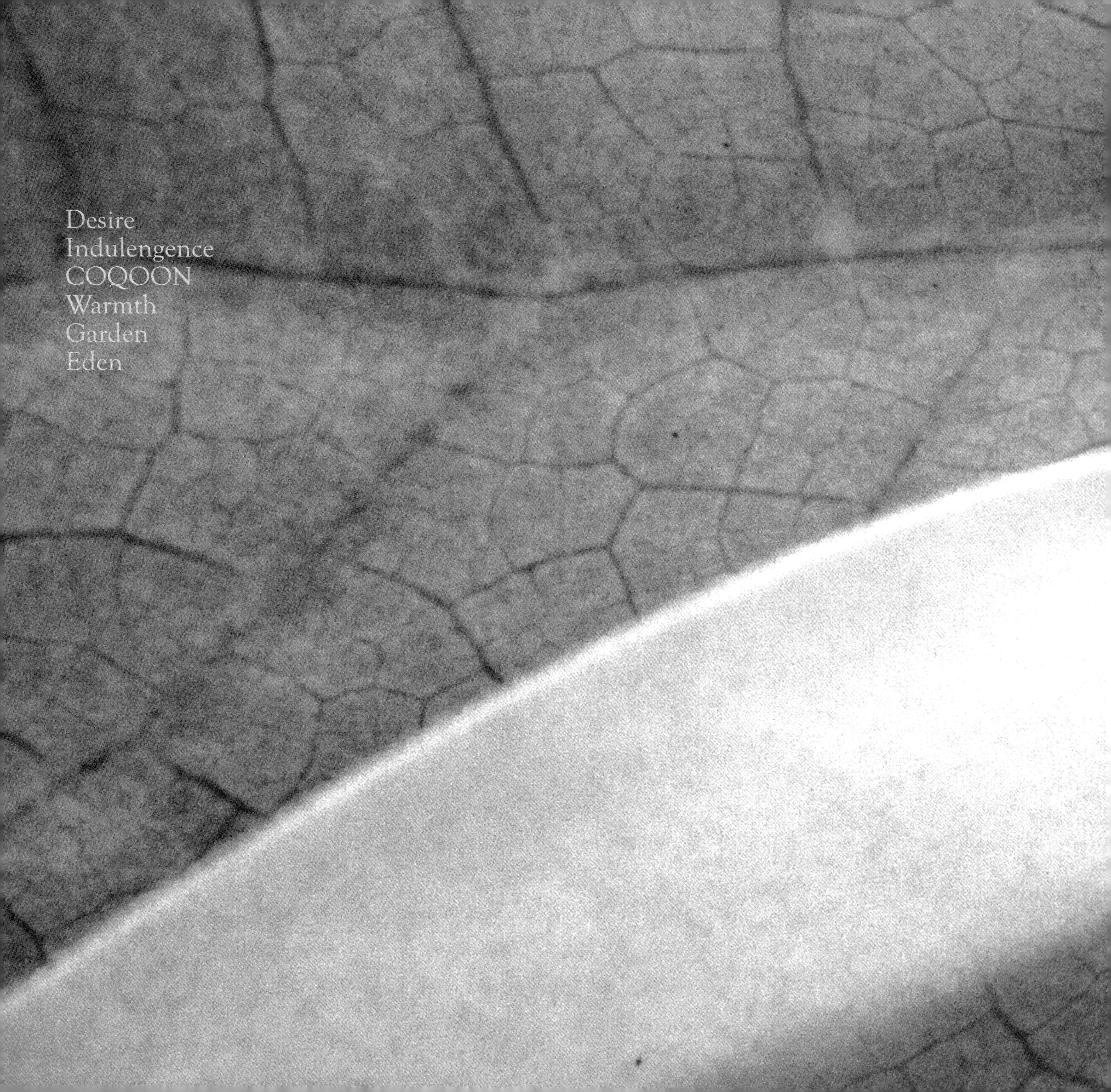

Desire
Indulengence
COQOON
Warmth
Garden
Eden

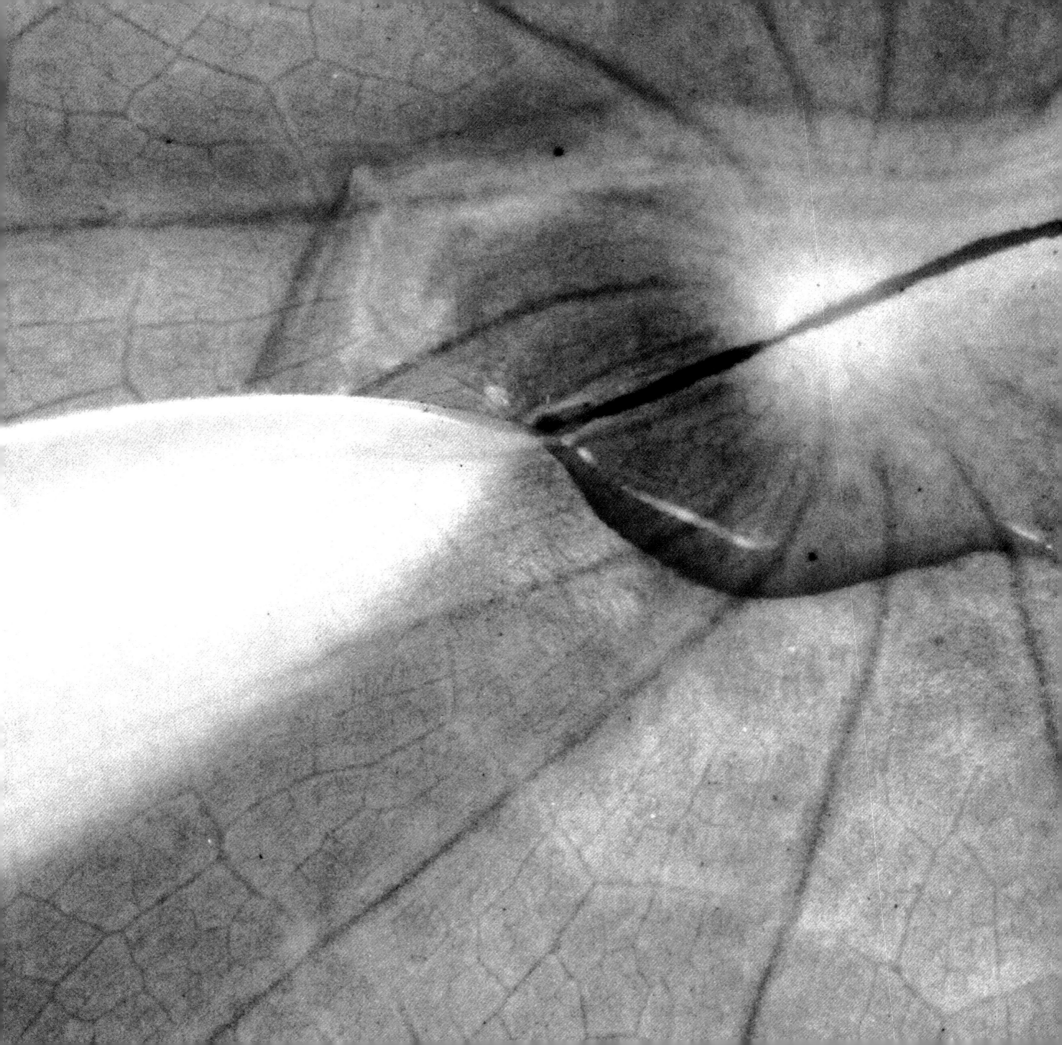

COQOON SPA

Indigo Pearl Phuket, Phuket

A nest surrounded by designer rattan and embraced by one of Phuket's oldest banyan tree.

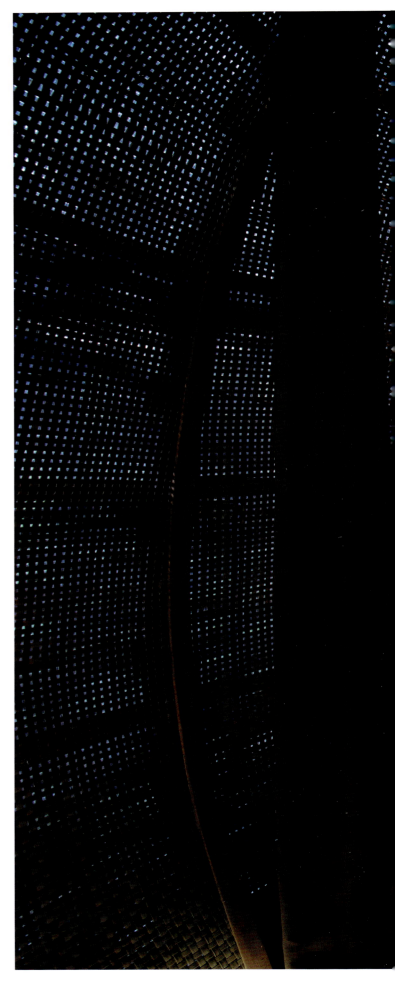

Everything slows down magically when you turn off the main road and enter the palm-lined driveway. Immediately, you begin to sense the tranquility of this magic place. But it's only when you walk into the heart of the spa that you realize how spacious the grounds are. And how absolutely magnificent!

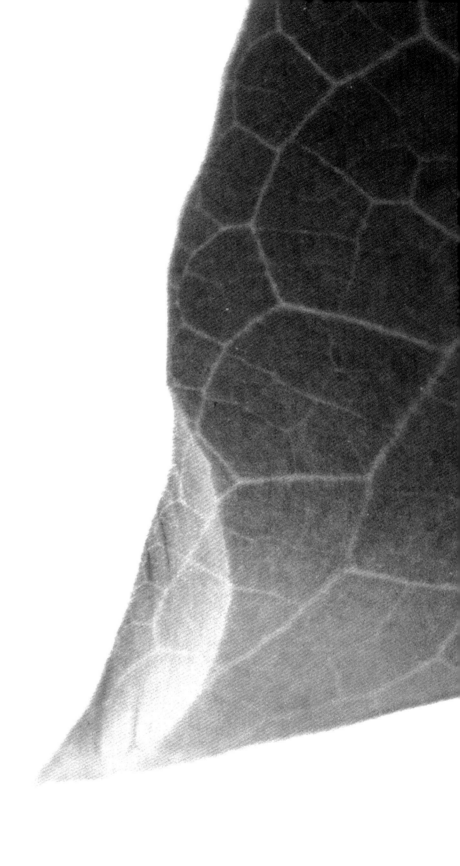

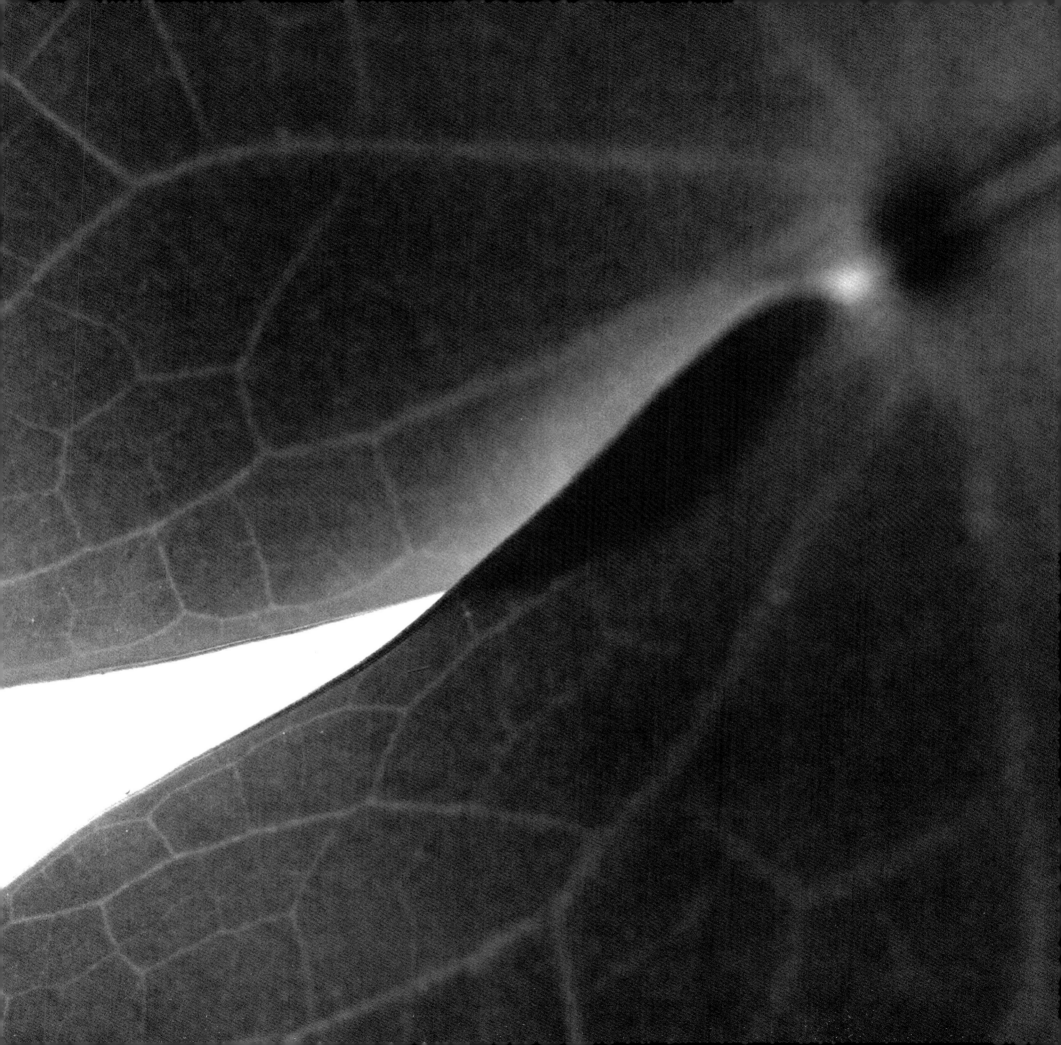

THE FOREST SPA

Tamarind Springs, Koh Samui

Private massage pavilions at Tamarind Springs designed for two, is a breezy wood and thatch structure - an open "sala suspended above the surrounding coconut grove. Lie back and be lulled by the sound of the wind sighing through the trees.

PROJECT DATA

Away Spa

Location
W Retreat Koh Samui
4/1 Moo 1, Maenam, Koh Samui,
Surat Thani 84330
T: +66 (0) 7791 5999 F: +66 (0) 7791 5998
E: whateverwhenever.wkohsamui@whotels.com
www.whotels.com/kohsamui

Design Firms
Interior Designer: P49 Deesign and Associates
Architect: MAPS Design Studio

W Retreat Koh Samui is about escape, escape into a world of relaxation intermixed with a world of wow! where understated cool meets warmth of space. It is the interplay of shadow with light, smooth form with texture, simplicity of colour with splashes of brilliance, and a place where the interior and exterior landscapes are inseparable. A key design direction was to maintain a ' Thai Zen' attitude throughout the Spa. Overall, this was achieved through clean lines, spaces that flow, design rhythm, and above all, the use of materials that symbolize relaxation and warmth.

The spa complex at W Samui is a masterpiece of architectural integration with the landscape and comprises a series of spa villas, relaxation spaces, the 'Tonic' juice bar and the wonderful and unique 'Thaimazcal' signature treatment facility. The spa villas are designed to be fully self-contained with an interior image based on total relaxation within the overall design strategy. Wood finishes play a significant role in this, as do the views of the sea beyond. The overall wood tone used throughout the spa was based on the bleached colour you see on driftwood that has been washed up on shore and exposed to the sun for a long time, essentially a grayed wood with a white undertone. This colour has been utilized for all interior wood based elements.

Another layer of design that enhances the 'W' attitude has been the integration of whimsical art and artifacts, continuous elements of surprise! Thought provoking, interesting, colourful and contrasting, these elements have been scattered throughout the spa.

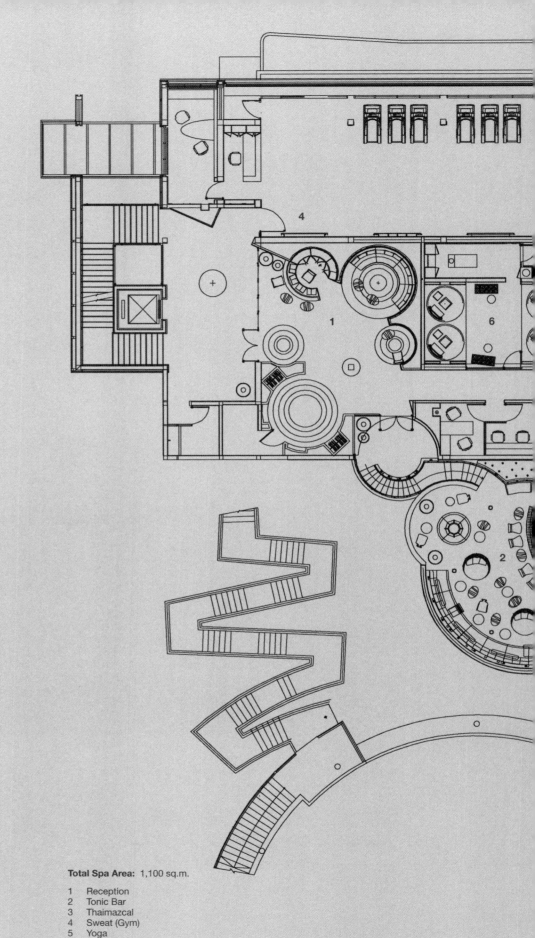

Total Spa Area: 1,100 sq.m.

1   Reception
2   Tonic Bar
3   Thaimazcal
4   Sweat (Gym)
5   Yoga
6   Salon
7   Changing Room
8   Single Suite
9   Suite
10 Deluxe Suite

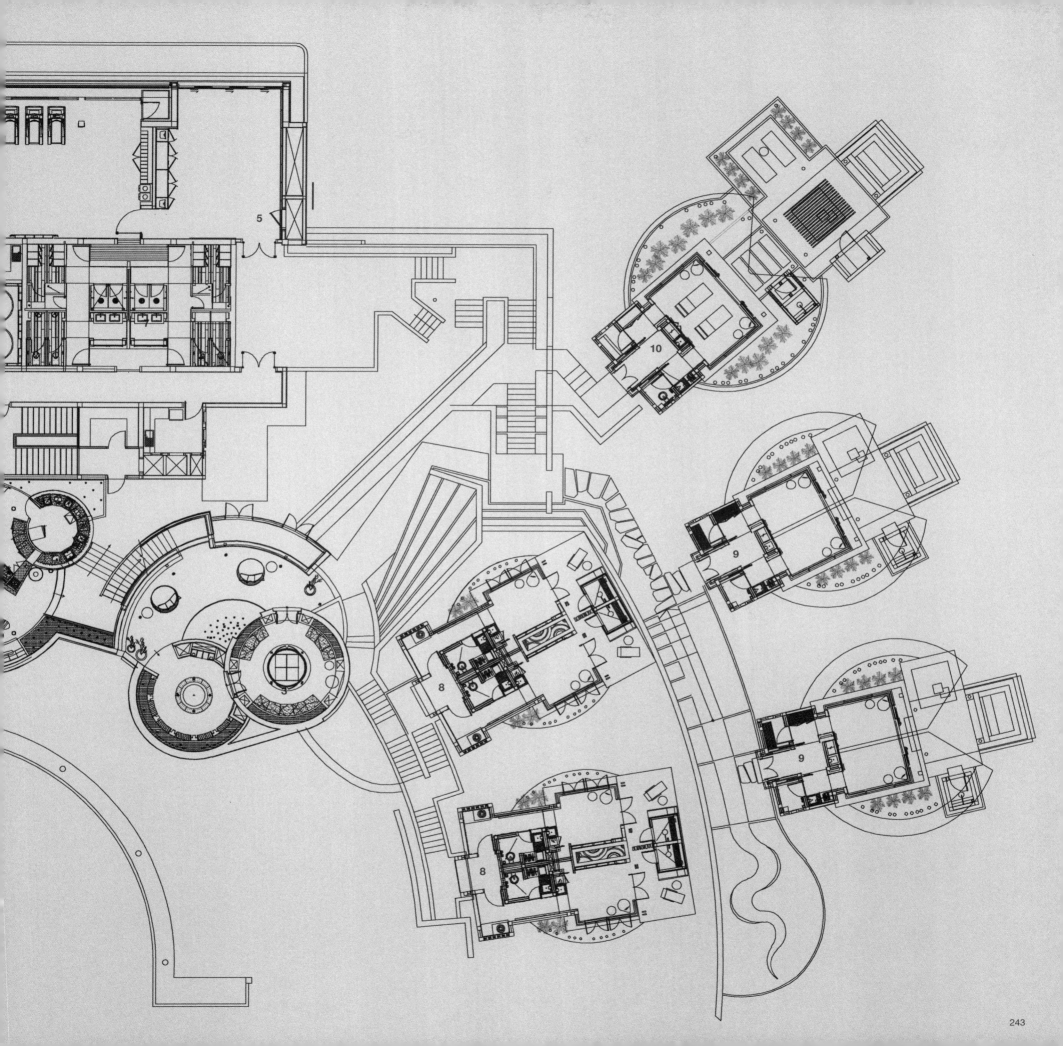

Elemis Spa

Location
The St. Regis Bangkok
159 Rajadamri Road, Lumpini, Bangkok 10300
T: +66 (0) 2207 7779
E: elemis.spa@stregis.com
www.elemisspabangkok.com

Design Firms
Interior Designer: Brennan Beer Gorman
Architect: Architects LLP, Plan Associates
Landscape Architect: Studioria

Elemis, the leading British professional spa and skincare brand, has opened an Elemis spa at the new St. Regis in Bangkok. The existing Elemis premises in London, Miami have been hailed as amongst the world's finest day-spas, and Elemis Bangkok is poised to follow suit. The menu comprises over twenty-five treatment options, all developed by Elemis and delivered by highly trained Elemis therapists using professional spa-strength Elemis formulations to achieve maximum results. The St. Regis Bangkok is ideally located on Rajadamri Road close to Bangkok's major business, shopping and entertainment areas and is just a five-minute walk from the famous Lumpini Park.

Elemis Spa at the St. Regis Bangkok is an extraordinary setting of Spa elegance that reflects the Elemis balance of tradition and innovation, offering a luxurious haven of rejuvenation from the hectic pace of Thailand's capital. Beyond pampering, guests of Elemis Spa will have their beauty and wellness concerns lavishly addressed with scientifically-proven treatments, lifestyle enhancing products and wellbeing traditions that encourage continuing transformation, all tailored to meet their individual needs.

The ultra-modern treatment areas are designed in soothing sand and cream tones with accents of soft white leather and silvery Thai silk throughout, and feature sophisticated skin and body care equipment. Elemis Spa features two levels of extensive amenities including water features and relaxation zones designed to help guests de-stress before treatments begin and to extend therapy benefits when treatments are completed. A mezzanine level of wet facilities includes Rassoul Chambers in which guests can experience an ancient Arabian bathing ritual. These signature facilities are further accompanied by steam rooms, hot and cold plunge pools and Experience Showers that allow for a personalized selection of water pressure, lighting and music. Each locker area features comfortable seating lounges, suspended Relaxation Pods that float like nests above the water features, providing a unique place to recline and rest. A Reflexology-inspired Foot Ritual Pool of smooth stones, triggers acupressure points, unblocks energy channels and improves overall feelings of wellbeing. In addition, a room with therapeutic lighting and music offers a private place for meditation, sleep and sound therapy to help balance energy levels.

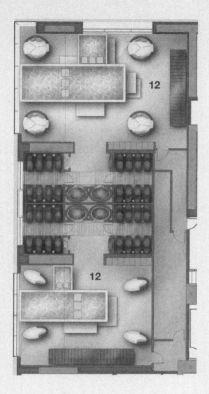

Upper Floor

Total Spa Area: 1,000 sq. m.

1   Reception
2   Double Treament Suite
3   Facial Treatment
4   Treatment Hydro
5   Facial Treatment
6   LPG Room
7   Body Treatment
8   Thai Double Treatment
9   Sensory Suite
10 Male Changing and Relaxation
11 Female Changing and Relaxation
12 Wet Facilities

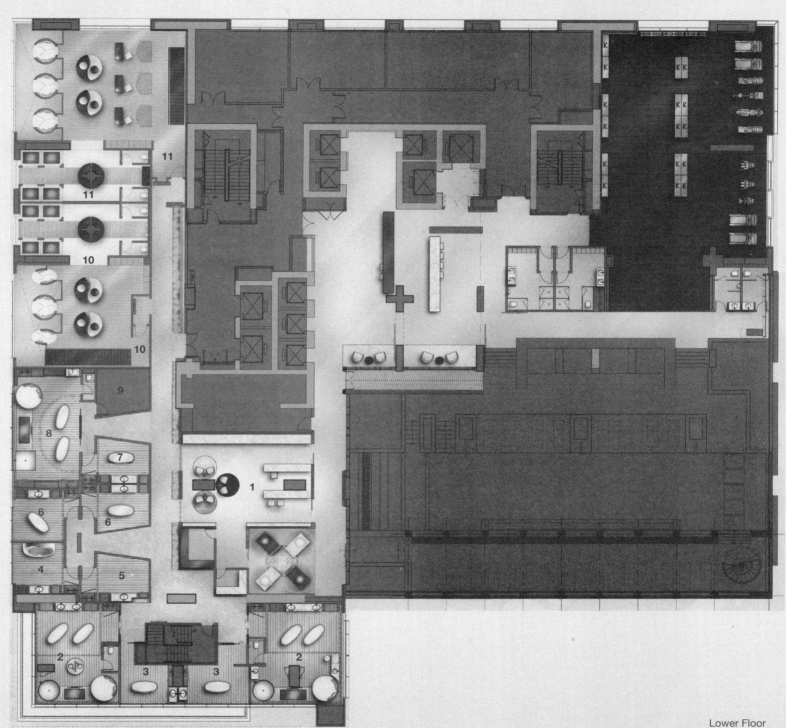

Lower Floor

eforea Spa at Hilton

Location
Hilton Pattaya
333/101 Moo 9 Nong Prue, Banglamung,
Pattaya, Chonburi 20260
T: + 66 (0) 3825 3000 **F:** +66 (0) 3825 3028
E: pattaya.info@hilton.com
www.pattaya.hilton.com

Design Firms
Interior Designer: August Design Consultant
Facade Design Architect: M.A.A.R
Landscape Architect: T.R.O.P

Eforea spa at the Hilton features an exclusive menu of treatment journeys and innovative design elements. It's promise is to bring balance and wellness to the body, taking guests on a transformative journey of the senses from which they emerge brighter, like a butterfly freed from its cocoon. On arrival guests are greeted with the warm hospitality for which Hilton is known. Guests can select from the wide variety of therapies offered in the nurturing and soothing environment. When participants emerge from eforea their bodies are renewed, their spirits lifted, and their outlook brightened. Since not every individual seeks the same path, eforea presents three distinct treatments, each comprising unique products and journeys designed exclusively for the Hilton. This journey of a thousand miles begins with a single step.

The design and the layout of the spa compliment the hotel's 'unique by design' philosophy. It boasts an impressive array of facilities including six private treatment pavilions containing state of the art equipment, personal changing areas, steam showers and private relaxation spaces. Lighting, which brightens and then softens, guides guests on their journey through walkways and ponds to the treatment pavilions. The spa's Thai massage pavilion embraces the wonderful Thai culture and offers a nurturing environment where guests can experience this highly specialized and therapeutic treatment. The Vichy Pavilion offers guests a cocooning hydrotherapy spa experience by combining the healing powers of water with the therapeutic benefits of massage in m, one heavenly experience. The interactive thermal experience allows guests to surrender their stress by embracing the warming elements of aroma infused steam and hydrotherapy massage, perfect to indulge in before or after your treatment experience. A salon provides hair and nail services in a warm and friendly environment and spa goers are treated to an uplifting tea, originating from a Thai family recipe of the hotel's spa manager. This special blend of orange, ginger and honey will immediately put all guests into a state of serene calm.

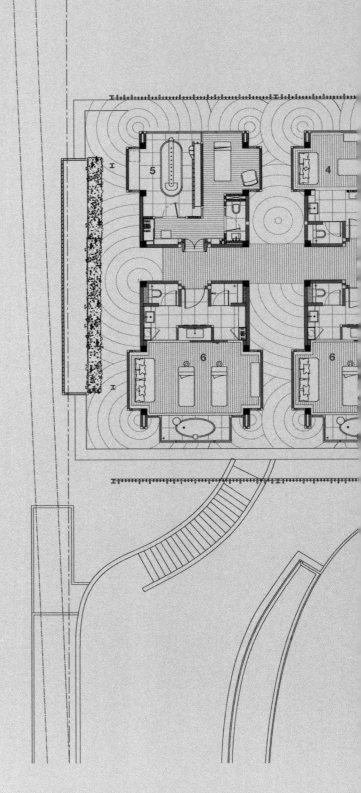

Total Spa Area: 880 sq.m.

1   Reception Area
2   Salon and Manicure/ Pedicure Room
3   Thai Pavilion
4   Single Treatment Room
5   Vichy Room
6   Couple Treatment Room
7   Thermal Room
8   Relaxation Area
9   Changing Room

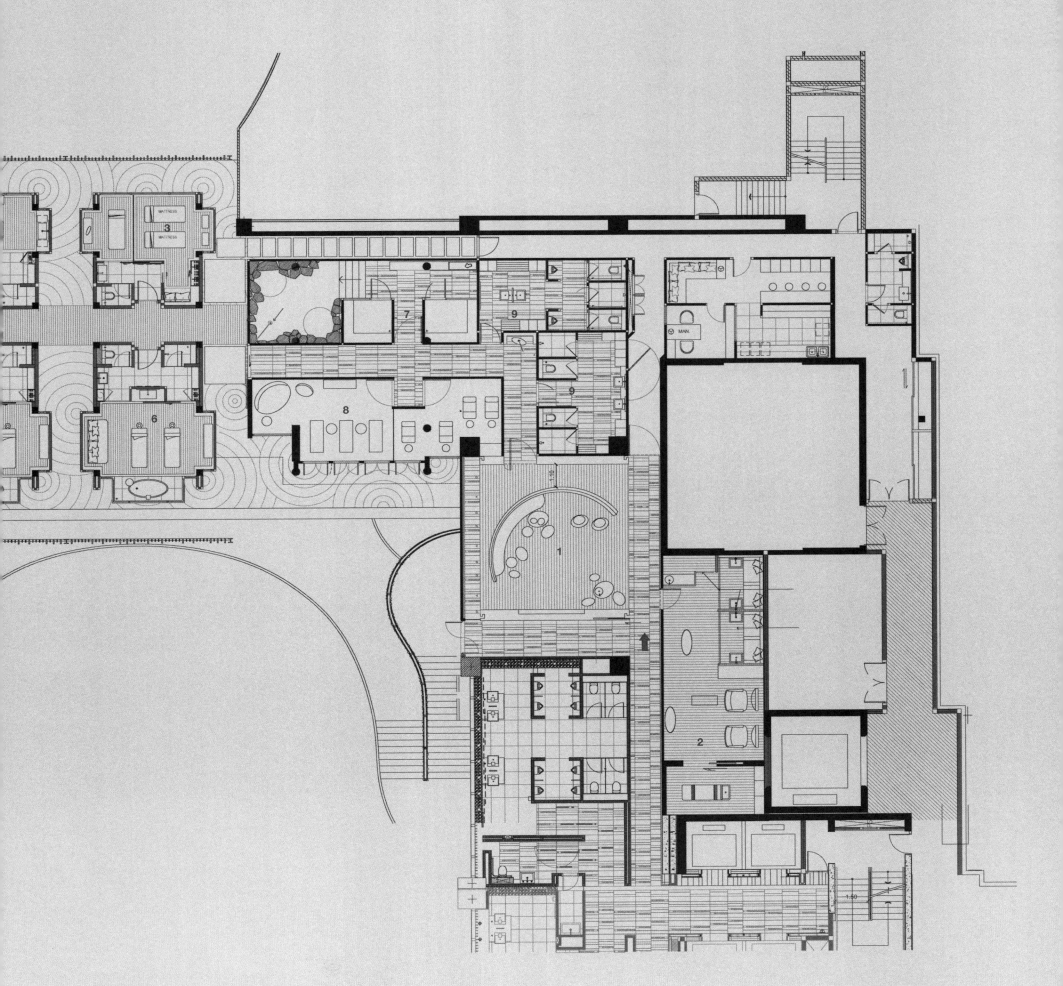

So Spa

Location
Sofitel So Bangkok
2 North Sathorn Road, Bangrak, Bangkok 10500
T: + 66 (0) 2624 0000 F: + 66 (0) 2624 0111
E: H6835@sofitel.com
www.sofitel.com

Design Firms
Interior Designer: PIA Interior
Architect: The Office of Bangkok Architects

The décor of Sofitel So Bangkok's So Spa is inspired by the legendary *Himmapan* forest in the Himalayas filled with mythical plants and fantastic heavenly creatures, some of them half-human, half-beast. To depict this magical forest invisible to human eyes, we created a mysterious light dappled experience where meandering paths and rich layers of natural scents resemble a mystical land worthy of exploration.

Overlooking the verdant Lumpini Park to the north and urban-scape to the south, the 630 sq.m. space is divided into the three primary functions; the reception and common area is the middle flanked by two wings.

Rustic finishes of the Urban Jungle reception frame views of the park below and the skyscrapers beyond. Here, guests can learn about spa products or sip healthy beverages at the Discovery Bar. A foot reflexology room and a relaxation room with four beds round out the common area.

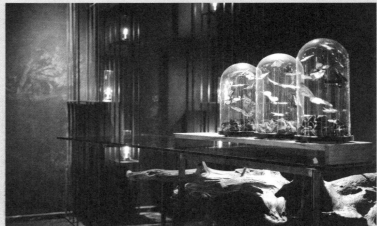

The East Wing houses seven massage rooms: one Thai massage room, four standard rooms, and two Massage Suites with bathtubs. The West Wing has three spacious Spa Suites consisting of a bedroom, bathroom and spa treatment room. PIA proposed this idea so that guests have the option of staying in the bedroom wing, or just using the massage room and bathroom. Since PIA's was only retained to design the public spaces, these are the only bedrooms in the hotel designed by them.

Unobtrusive black marble flooring throughout allow eyes to focus on rough wood elements such as the salvaged floor to ceiling-height teak trunks along corridors and common areas, and the scratched coarse-grained *Jamjuree* (rain tree) wall paneling on the reception's walls. Wood latticework ceilings in the common area create an interplay of light and shadow along walls and floors.

Wall treatments throughout the spa are drawn from traditional temple murals. Prints lifted from temple depictions of *Himmapan* forest scenes are translated into black and white wallpaper for corridors, with additional details painted on by a local artist. In massage rooms, stipple-effect stylized designs on perforated backlit aluminum panels resembling a 3-D forest glade serve as decorative lighting.

It is hoped that guests will leave the spa feeling refreshed and uplifted as if woken from a magical slumber.

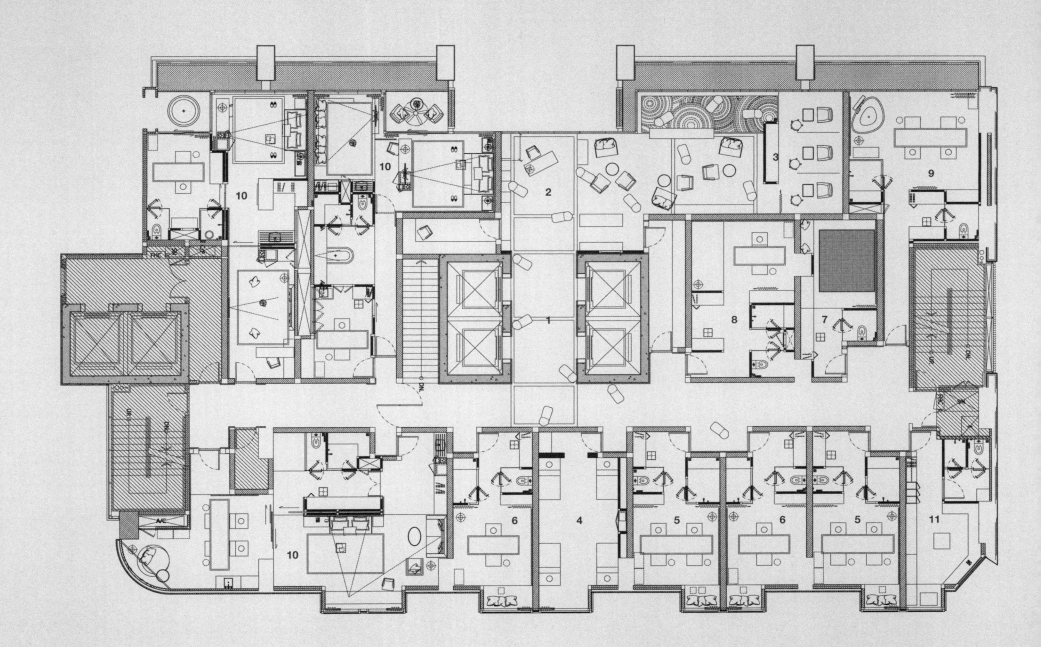

Total Spa Area: 630 sq. m.

1   Lift Corridor
2   Reception
3   Foot + Hand Massage
4   Relaxation Area
5   Double Massage Room
6   Single Massage Room
7   Thai Massage Room
8   Single Massage + Tub Room
9   Double Massage + Tub Room
10 Suite
11 Staff Room

Kempinski The Spa

Location
Siam Kempinski Hotel Bangkok
991/9, Rama I Road, Pathumwan, Bangkok 10330
T: +66(0) 2162 9050 F: +66(0) 2162 9301
E: spa.siambangkok@kempinski.com
www.kempinski.com

Design Firms
Interior Designer: Hirsch Bedner Associates
Architect: Tandem Architects 2001
Lighting Designer: Intaran Design

Kempinski The Spa is located on the 7th floor of Siam Kempinski Hotel Bangkok. The property has been built on what is historically a very important part of the city, the site having once been part of the Lotus Pond Palace commissioned by King Rama IV (known in the West from the story *The King and I*) as a retreat outside the city. The palace gardens were famous for their beauty and serenity, and were home to many rare plants and flowers, as well as birds.

With its aim to create a peaceful sanctuary, the property is built in a circle with lush landscaped tropical gardens and three salt-water pools at its centre, thus recreating the historic palace gardens.

Throughout the hotel, the theme of lotus flowers and splashing water plays an important role in reflecting the area's history. Lotuses can be found in the designs of the spa, in prints and photos in the treatment rooms, as well as in the fresh flower decoration. Furthermore, purple, one of the shades of the lotus, is the principal colour used throughout the property.

Another important element in the design concept is the artwork. Part of the hotel, including the spa, comprises an extensive collection of Thai art, which blends harmoniously into the overall architecture and design of the property. Nineteen professors and students of the College of Fine Arts were commissioned to create paintings and sculptures especially for the hotel.

In Kempinski The Spa, there are some outstanding artworks installed in the space. Guests are greeted as soon as the lift opens by a black sculpture with water and fresh lotus floating on the surface. In the relaxation room, two round sculptures are aesthetically placed in front of a textured wall to create a calm, Zen-like atmosphere. The door handles of each treatment room are made from cast brass shaped in lotus forms, so emphasizing the décor theme. On entering the room, all eyes are attracted to a magnificent image of the lotus on the wall. This looks as if carved on sandstone, although it is actually hand-made from mulberry paper.

These artworks add elegance to the refined atmosphere of Kempinski The Spa, while the beauty of the space is further enhanced by the widespread use of natural materials, wood, mulberry paper, sandstone, cast brass, and ceramics.

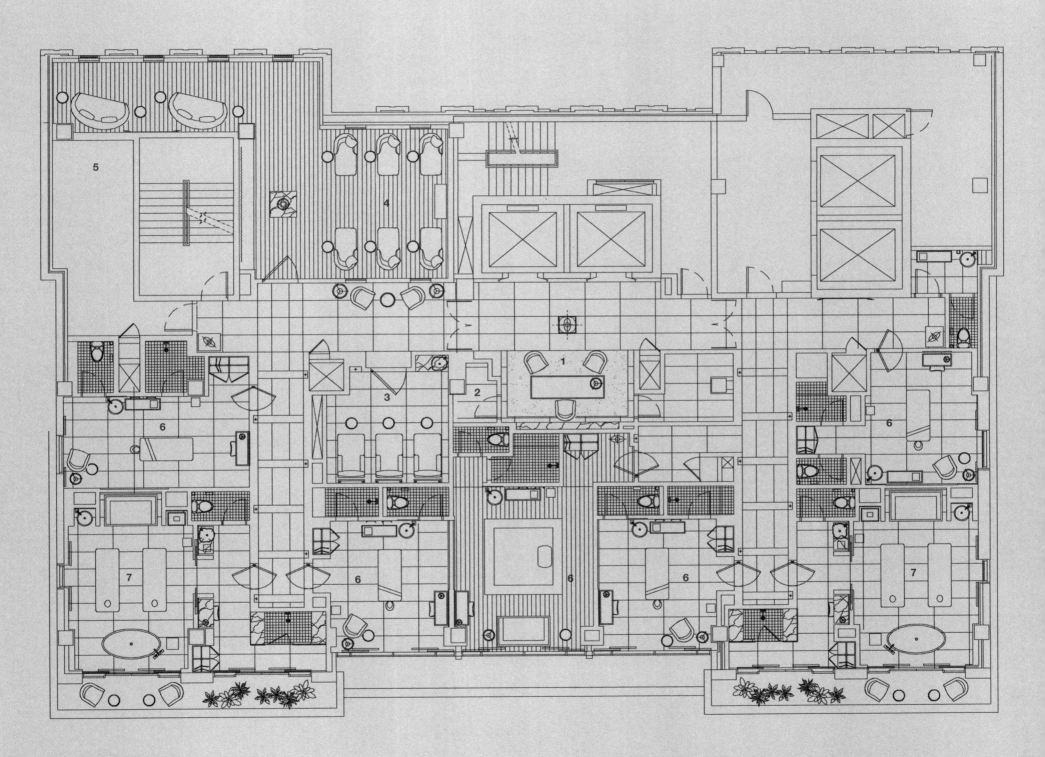

Total Spa Area: 500 sq.m.

1   Reception
2   Store
3   Foot Spa
4   Relaxation Area
5   Service Bar
6   Treatment Room
7   Treatment Room (Double)

Devarana Spa

Location
Dusit Thani Bangkok
946 Rama IV Road, Bangkok 10500
T: +66 (0) 2636 3596 F: +66 (0) 2636 3597
E: bangkok@devaranaspa.com
www.devaranaspa.com

Design Firms
Conceptual Design: Ploy Chariyaves
Interior Designer: PIA interior

Devarana Spa offers an atmosphere of ultimate relaxation in its very own healing garden of heaven. With an emphasis on wellness and 'East meets West', Thai health and beauty practices have been sourced from age-old therapies then updated with modern knowledge to pamper and revitalize guests.

The term 'Devarana' [pronounced té-wa-run] is derived from Thai-Sanskrit and means 'garden in heaven'. The term dates back to ancient Thai literature called 'Traiphuum Phra Ruang' written by Phraya Lithai. In this literary work, the writer described this particular place as situated at heaven's gate with a heavenly scent and nurturing environment, surrounded by gardens and ponds. The garden's décor glimmers with silver and gold, and is filled with natural stones and gems. In the garden, people can hear soft, melodic music played by harp, flute, and other traditional musical instruments.

Inspired by this concept of the garden in heaven, Devarana Spa excites the senses and makes guests feel truly special. Luxurious pampering and healing treatments are offered in a soothing, stress-relieving environment. The contemporary Thai design and décor invites guests to transcend the everyday world. The spa provides a relaxing and welcoming atmosphere in which to enjoy premium spa treatments with traditional Thai service and hospitality. The concept for the spa was design by one of Thailand's renowned writers, Ploy Chariyaves.

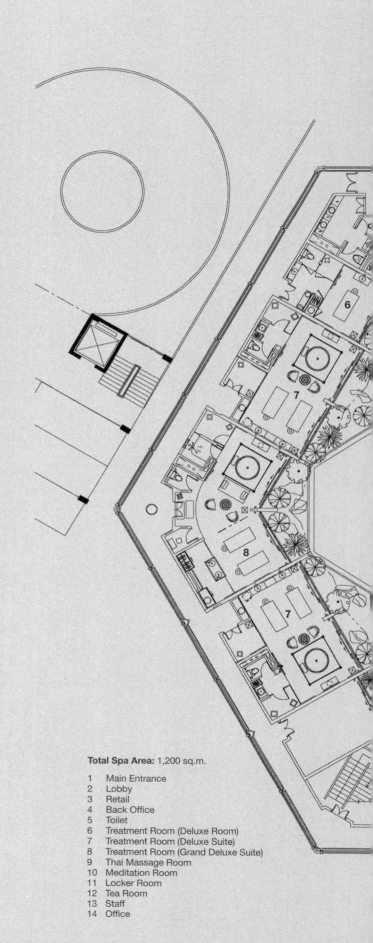

Total Spa Area: 1,200 sq.m.

1 Main Entrance
2 Lobby
3 Retail
4 Back Office
5 Toilet
6 Treatment Room (Deluxe Room)
7 Treatment Room (Deluxe Suite)
8 Treatment Room (Grand Deluxe Suite)
9 Thai Massage Room
10 Meditation Room
11 Locker Room
12 Tea Room
13 Staff
14 Office

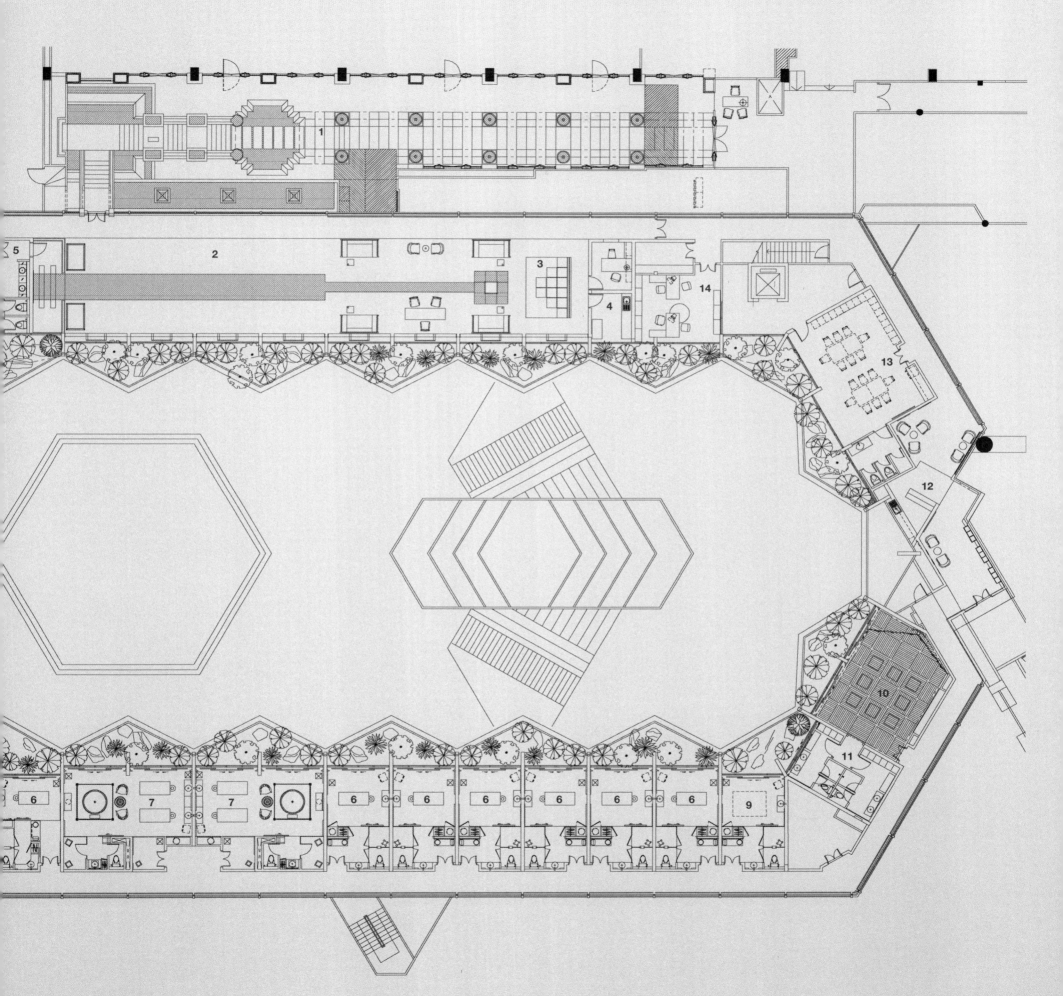

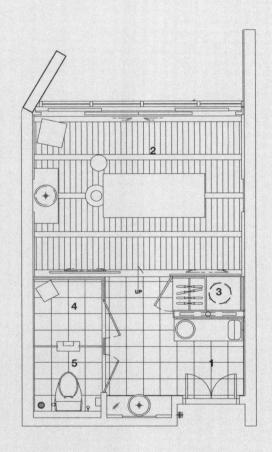

Deluxe Room

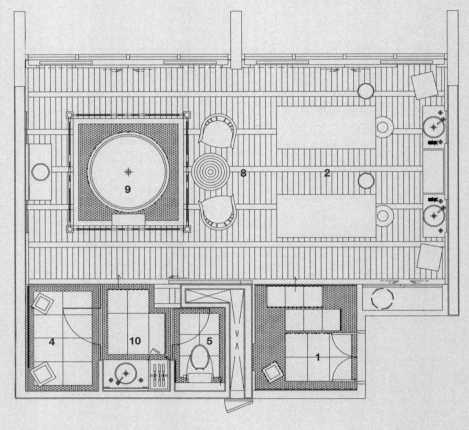

Deluxe Suite

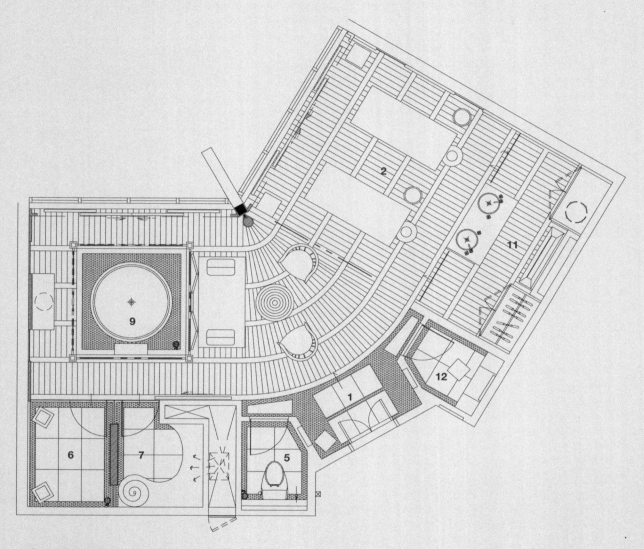

1   Foyer
2   Treatment Area
3   Clothes Area
4   Shower & Steam Room
5   Toilet
6   Shower Room
7   Steam Room
8   Relax Area
9   Bath Tub
10 Changing Area
11 Wash Basin & Closet
12 Dressing Room

Grand Duluxe Suite

Thann Sanctuary

Location
Crowne Plaza Phuket
8/88 Moo 7, Sakdidech Road, Vichit, Phuket 83000
T: +66 (0) 7630 2900 ext. 6143
F: +66 (0) 7630 2926
E: stay@crowneplazaphuket.com
www.crowneplaza.com

Design Firms
Interior Designer: PIA Interior
Architect: Surapat Architect
Lighting Designer: Withlight

The spa is designed with the concept 'Lucid Sanctuary': a calm, tranquil and soothing retreat hidden deep within the hotel that serves as a spiritual escape from the rest of the world. The sanctified space uses dark and mystical undertones to play with the emotions on a deeper level. It draws on rustic elements used in the lobby, such as the teak wood latticework that wraps around the walls and ceilings like a textured blanket, and the smooth Travertine floors that contrast with the rough rugged slate walls. The dramatic eerie yet soothing lighting, set against the palette of contrasting rustic and natural materials and linear geometry, creates a mysterious spa experience.

The hotel and spa draw design inspiration from the warmness of shade and shadow. To discover this spacious spa, one begins with the stunning entrance tunnel covered with layers of vertical and horizontal wooden strips leading to the open-pan corridor that connects to the spa reception. These are continued in ponds along the corridors, and this gives a characteristic sense of flawless and natural continuity.

The lobby is furnished with fine tropical wood pieces formed as a weaving pattern to create shade and shadow that flow through the entire space. The atmosphere is augmented by contemporary designs using warm materials carefully handpicked to create visual interest. These create the shading effect of a modern sanctuary. Integrating straight wooden lines with handcrafted finishes and unique art pieces works to interpret the combined wisdom of traditional massage techniques and contemporary scientific approaches.

The main corridor continues from the lobby to the seven exclusive treatment rooms that include four single bed treatment rooms and three double bed treatment rooms. This corridor is lined with ponds and has walls decorated with rattan art pieces inspired by seashells. The treatment rooms design reflect a harmonious blend of the traditional Thai way and modern luxury. This effects a collision of wooden floors and black stone tile furnishings. Each room is decorated with local Thai artifacts from many regions such as traditional palm leaf fans and Thai silk framed in modern graphic compositions. All treatment rooms feature a private balcony and expansive outdoor and indoor shower areas connected with a Jacuzzi picturing views of the lush garden. Function and aesthetic are combined to enhance the serenity of true relaxation.

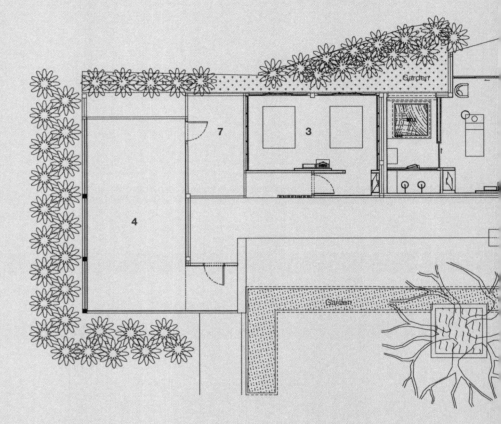

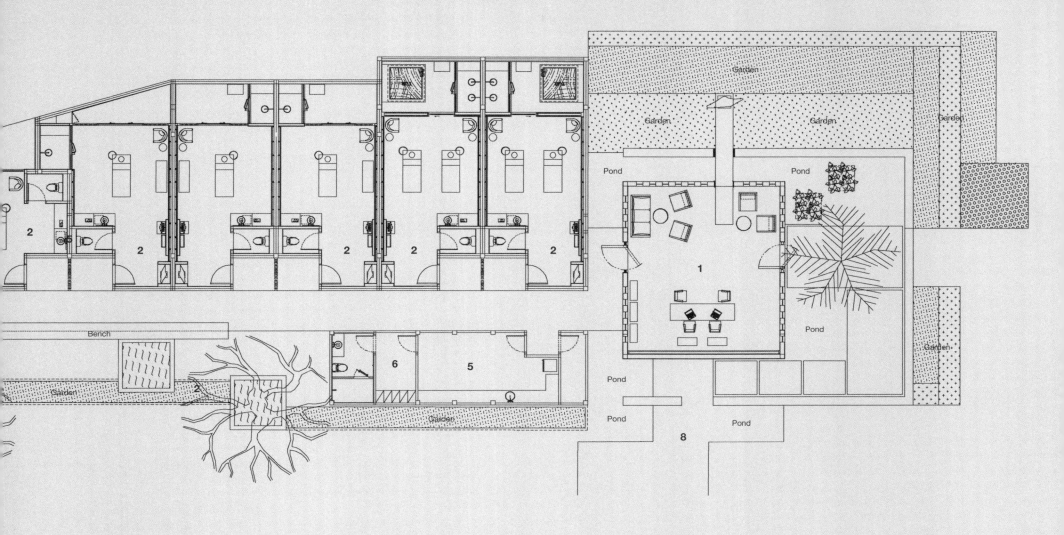

Total Spa Area: 1,000 sq.m.

1 Reception Area
2 Treatment Room
3 Massage Room
4 Yoga Room
5 Pantry & Preparation Room
6 Locker Room
7 Store
8 Guest Entry from Guest Room

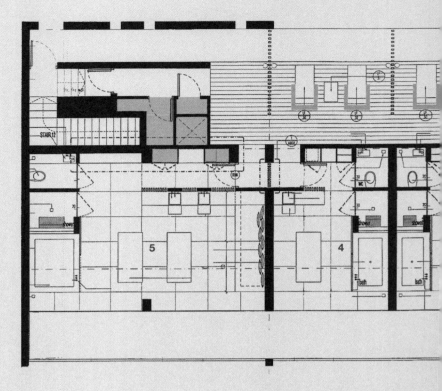

The Spa

Location
The Chedi Chiang Mai
123 Charoen Prathet Chang Khlan,
Muang Chiang Mai, Chiang Mai 50100
T: +66 (0) 5325 3333   F: +66 (0) 5325 3392
E: spa@chedi-chiangmai.com
www.ghmhotels.com

Design Firms
Interior Designer: Kerry Hill
Architect: Kerry Hill, Tandem Architects 2001
Landscape Architect: Kerry Hill

Distinctive teak doors lead from the spa's strikingly minimalist exterior to an inner sanctum where the emphasis is on pure holistic pampering. Only the finest products are used, sourced from far and wide. Post-treatment, guest may opt to linger on the rooftop deck overlooking the Mae Ping or join sunbathers below on cosy loungers that line the 34-metre swimming pool.

Located adjacent to the swimming pool, the spa facilities include ten private, contemporary Asian style spa suites, complete with steam showers and terrazzo bathtubs. The deluxe double spa suites are equipped with herbal steam rooms, aromatic sauna and terrazzo bathtubs. In addition, the manicure and pedicure lounge offers a full range of beauty treatments. The retail section offers a wide selection of spa and beauty products that have been specially created for The Spa at The Chedi.

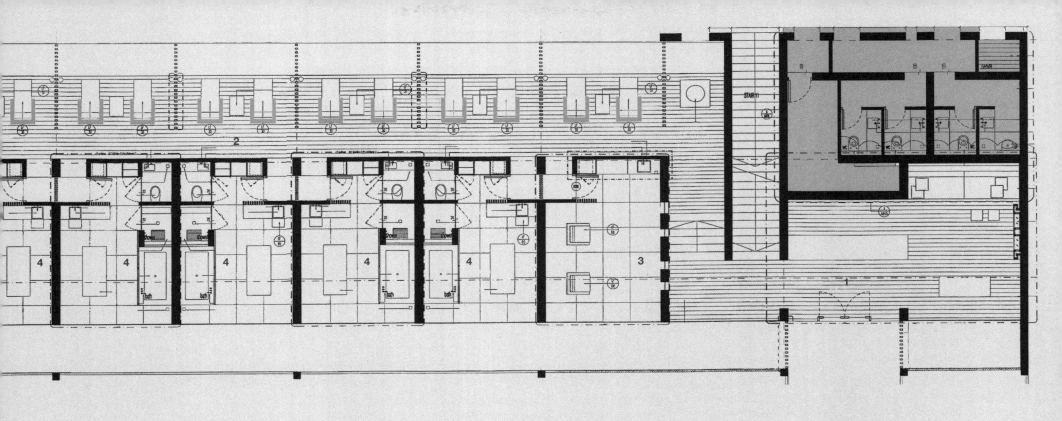

Total Spa Area: 1,109 sq.m.

1 Spa Reception
2 Spa Lounge
3 Manicure & Pedicure Room
4 Treatment Room
5 Treatment Room Double

Anantara Spa

Location
Anantara Phuket VIllas
888 Moo 3, Mai Khao, Thalang, Phuket 83110
T: +66 (0) 7633 6100-9 F: +66 (0) 7633 6177
E: phuket@anantara.com
www. spa.anantara.com

Design Firms
Interior Designer: P49 Deesign and Associates
Architect: Habita Architects
Landscape Architect: Bensley Design Studios

As the resort is located in one of Thailand's popular beach destinations, the Had Mai Khoa beach in Phuket, the designer of the Anantara Spa adopted the architectural forms of this rural tourist destination as a driving concept for the interior design. The layout was then aligned with the architecture and integrated with the resort's landscape architecture. Whether through the planned layout of furniture or the use of outdoor environments, the plan creates a fully connected interplay of interior and exterior worlds.

The lobby's reception area is focused on the principle axis leading to the featured partition behind the reception desk. This featured partition is decorated with woods adopting the rural patterns and rhythm that create a spectacular statement in the space. The casual seating layout and the home-sized furniture offer guests the warm and welcoming atmosphere of a rural Thai house.

The design concept for the library led to the creation of a living room with collections of antique items, furniture and appliances displayed throughout. The rhythmic repetition of the modern pattern on the featured white wall and the use of an antique bell as a chandelier to decorate the ceiling help raise the room's design appeal. The meeting room and boardroom are beautifully designed and employ the skillful placement of walls to separate the areas. The wooden pattern in the lobby is also used on the doors in this area, serving as a decorative wall. While this creates continuity with the design of the lobby, the two are differentiated through the use of graphic accents in the carpets' modern colours and patterns.

The reception area of the spa complex was also based on the same concept as the lobby. The designer made use of a gimmick found in street peddling, applying the technique to the display of spa products by transforming Thai-style food cupboards into consoles. The walls in the spa complex were painted betel-nut red to symbolize Thainess as well as to differentiate this area from other public spaces. Still, works of rural Thai folk art are retained in the form of furniture and decorative items.

The general layout of the pool villa plan and furniture emphasize the provision of spectacular views from nearly every angle. This responded to the architectural concept for guests of the pool villa which was to connect the external environment with the interiors as much as possible.

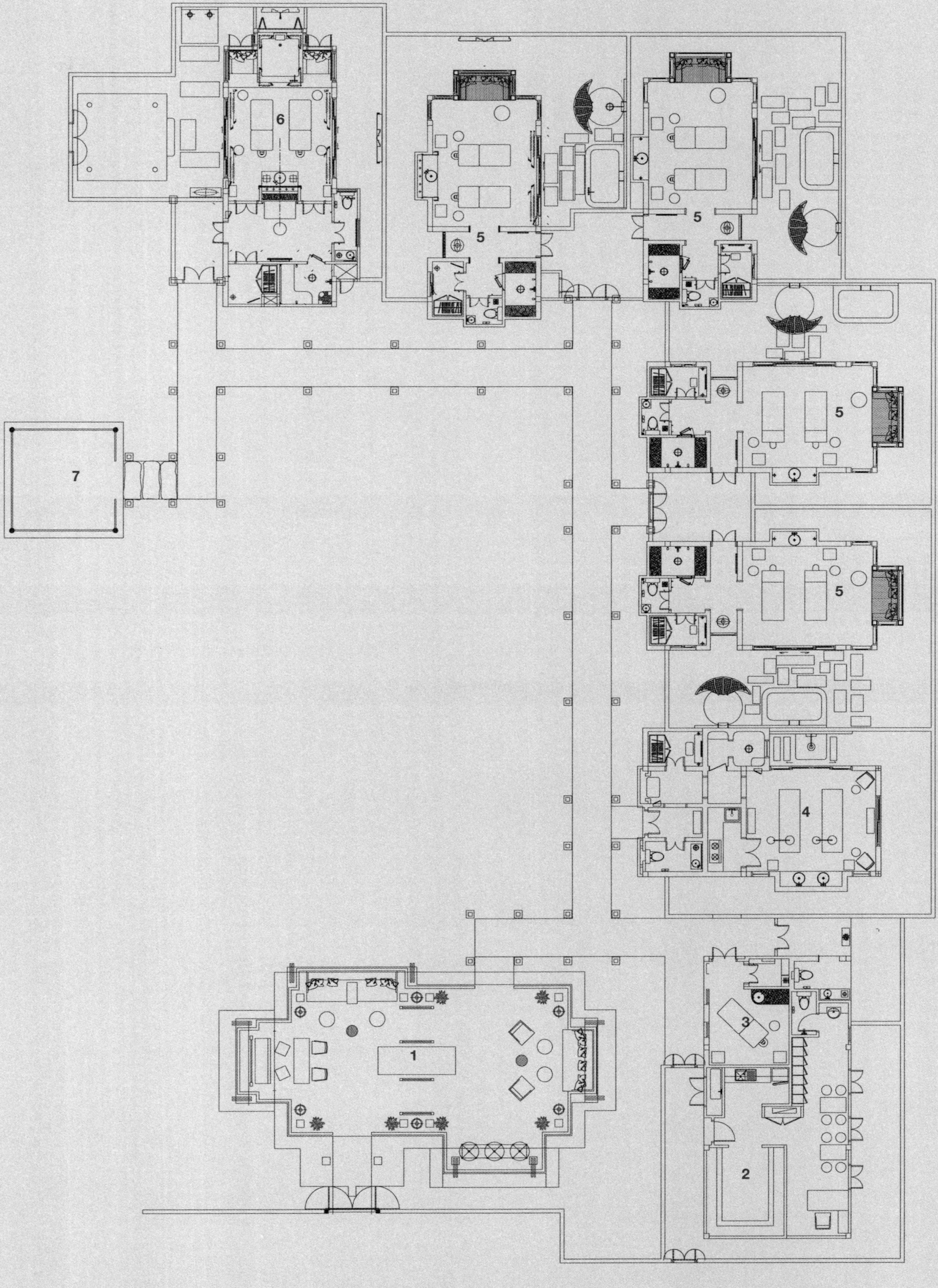

Total Spa Area: 400 sq.m.

1   Spa Reception
2   Back of House
3   Facial Treatment Room
4   Ayurvedic Room
5   Double Treatment Room
6   Suite Treatment Room
7   Sala

The Spa

Location
Four Season Resort, Chiang Mai
Mae Rim-Samoeng, Old Road, Mae Rim,
Chiang Mai 50180
T: +66 (0) 5329 8181   F: +66 (0) 5329 8190
E: reservations.thailand@fourseasons.com
www.fourseasons.com/chiangmai/spa/

Design Firms
Architect and Interior Designer: Bunnag Architects
International
Consultants
Landscape Architect: Bensley Design Studios

With the designer's deep passion for Lanna cultural heritage and architecture, the Four Seasons Resort spa in Chiang Mai is designed to match these ancient arts and nature. The master plan of the Four Seasons Resort is designed to place Lanna architecture amongst the rice fields and surrounded by dense teak trees, and architectural setting and style that has never appeared in Lanna history. Each building is located around the rice field and forms a U shape; the open space of the U heads toward the mountain in the west so sunlight creates the master plan silhouette of the mountain range, thrown into light and shadow from the side, and seen when looking out across the rice field. This idea is inspired by the plan of the old city of Chiang Mai that allows visitors to see the beautiful scenic mountain range in silhouette with the sunset behind when looking out from the old city.

The architecture of the living pavilion draws its inspiration from the bell pavilion of Wat Saenfang in Chiang Mai. These living pavilions are elegantly situated in the descending order of the hill skyline, complimenting the natural scenic beauty of the terraced rice fields and teak forests.

Significant Lanna decorative arts, such as gold painting from Viharn Numtam, corbels from several Lanna temples, Wat Ton Kwen's wooden carvings, and Wat Phumin's snake sculptures, are carefully selected and are directly employed with minor or minimum or adaptation. This reflects the designer's desire to preserve original Lanna cultural heritage.

The value of the architecture at the Four Seasons Resort lies in the blending of natural resources with cultural heritage. The aesthetic elements of ancient architecture are applied with an intention to create contemporary style not only for a commercial purpose but also with enduring social and cultural benefits.

Total Spa Area: 836.13 sq.m.

1   Private Garden Entrance to
    Spa Treatment Room
2   Treatment Room
3   Private Plunge Pool
4   Private Bath Tub Pavilion
5   Vichy Room

Entrance

The Dheva Spa

Location
Mandarin Oriental Dhara Dhevi Chiang Mai
51/4 Chiang Mai – Sankampaeng Road, Moo 1,
Tasala, Muang, Chiang Mai 50000
T: +66 (0) 5388 8888 Ext 8983
F: +66 (0) 5388 8999
E: mocnx-spa@mohg.com
www.mandarinoriental.com/chiangmai

Design Firms
Interior Designer: Lanfaa Devahstin
Architect: Aesthetics Architects
Landscape Architect: Suthat Phothisakha
Lighting Designer: Withlight

The Dheva Spa and Wellness Centre, in the secluded grounds of Mandarin Oriental Dhara Dhevi, offers a hedonist's haven: Thailand's first world-class destination spa, unique both in concept and architecture.

The Dheva Spa and Wellness Centre is a palatial 3,100 square metre sanctuary embellished with ornate moldings and sculptures depicting sacred animals and symbolic Buddhist motifs, loyally recreated by one hundred and fifty Chiang Mai artisans following an original Burmese template from Mandalay, Myanmar. The whole structure took these artisans over three and a half years to complete.

The spectacular exterior is matched by the sumptuous interiors where guests enjoy impeccable service and a range of treatments drawn from three continents, their origins spanning over four millennia. At the same time, The Dheva Spa also draws on time-honoured rites that are very much part of the ancient Lanna Kingdom, of which Chiang Mai was the royal capital. The tranquil treatment rooms, with Thai silks, polished teak wood and marble tiles, are punctuated with authentic Asian antiques.

The eighteen treatment rooms and suites have been designed for both individuals and couples, as a soothing refuge from urban stress and angst and a place where the journey to well being can begin. The Treatments at The Dheva Spa and Wellness Centre proudly offers a broad variety of European, Asian and North African relaxation therapies as well Indian holistic treatments based on Ayurveda, from the Sanskrit meaning 'science of life'. Among its numerous unique features there are soothing Vichy showers, elegant Watsu Pool and giant hammams beautifully appointed among the spa's dark wood panelled chambers. Popular Thai, Swedish and ancient Indian massage techniques are available, as well as a blissful range of aromatherapy treatments.

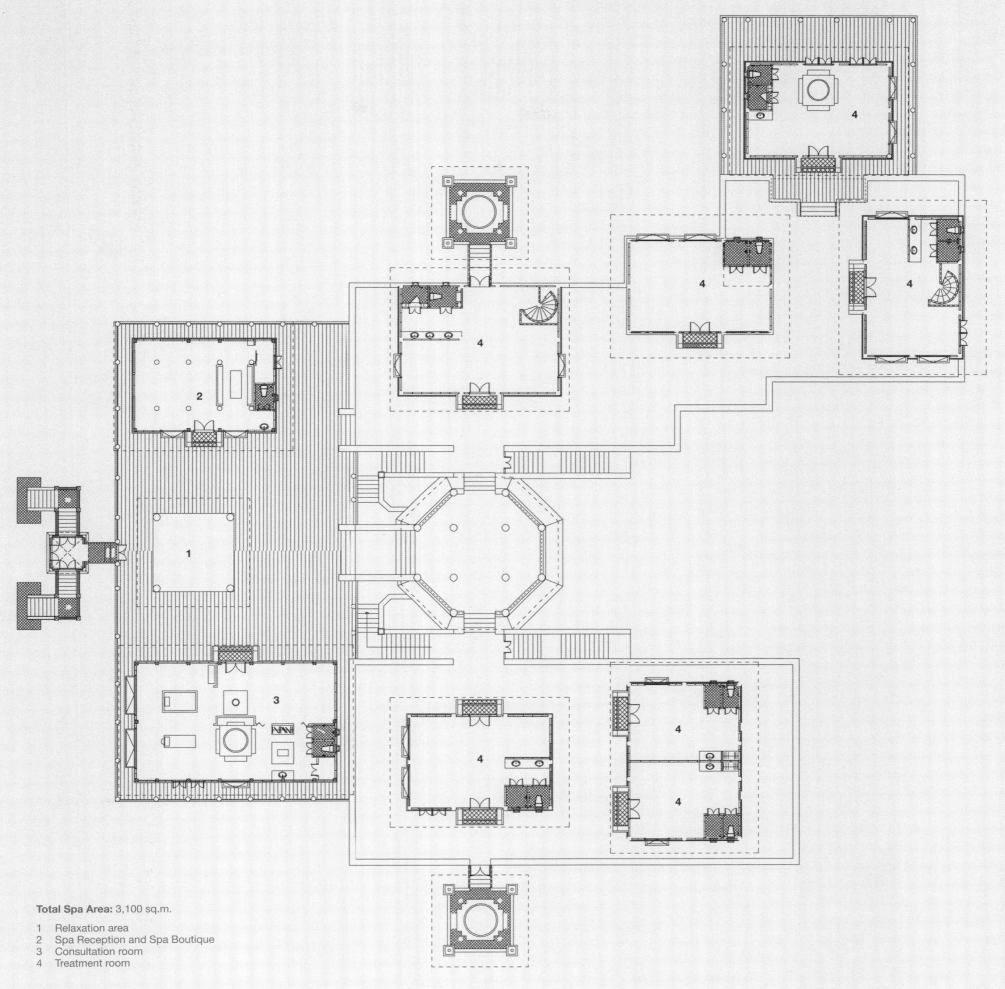

Total Spa Area: 3,100 sq.m.

1 Relaxation area
2 Spa Reception and Spa Boutique
3 Consultation room
4 Treatment room

ESPA

Location
Phulay Bay, A Ritz-Carlton Reserve
111 Moo 3 Nongthalay, Muang Krabi 81000
T: +66 (0) 7562 8111   F: +66 (0) 7562 8100
E: rc.kbvrz.spa.manager@ritzcarlton.com
www.phulay-bay.com

Design Firms
Interior Designer: Interior Architects 49
Architect: Architects 49 Phuket
Landscape Architect: P Landscape

ESPA at Phulay Bay, A Ritz-Carlton Reserve, is situated on the edge of the lagoon amid lush gardens and indigenous forest. The spa is a luxurious and indulgent wellbeing sanctuary for relaxation and rejuvenation. The contemporary spa is split-level and comprised of three traditional Thai pavilions offering eleven luxurious treatment rooms. These include singles and couples Thai massage rooms, two couples suites and two private spa suites with a relaxation pavilion overlooking the tranquil lagoon. The spa environment is soothing and the stylish setting perfectly captures the spirit of Ritz-Carlton Reserve.

ESPA provides luxurious thermal suites, and includes aroma steam room, sauna, vitality pools and a relaxation lounge. The gracious spa personnel will guide guests through these areas and then allow for a private discovery of the enchanting character of the spa facilities. The thermal suite provides a variety of warming, cooling and water experiences designed specifically to prepare the mind and body for the treatment experience to come.

The spa concept has been designed to assist guests in achieving a sense of balance; helping to consider lifestyle awareness in the quest for ultimate wellness. This understanding can be achieved with a wellness program based on a personal consultation that helps create a treatment and nutritional and alternative fitness routine for the duration, or part of, the guest's stay at Phulay Bay. Pilates, yoga and meditation classes can be integrated into the wellness program and a selection of other activities, such as trekking in the National Forest, can create a connection to the local natural beauty to deliver a sense of place and a truly memorable experience.

In addition to the eleven treatment rooms that include two VIP couple suites, the ESPA houses a café offering healthy juices and raw food, and a fully equipped yoga and meditation room.

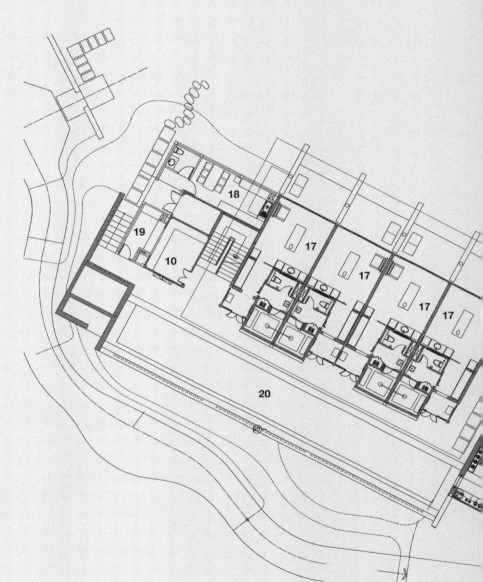

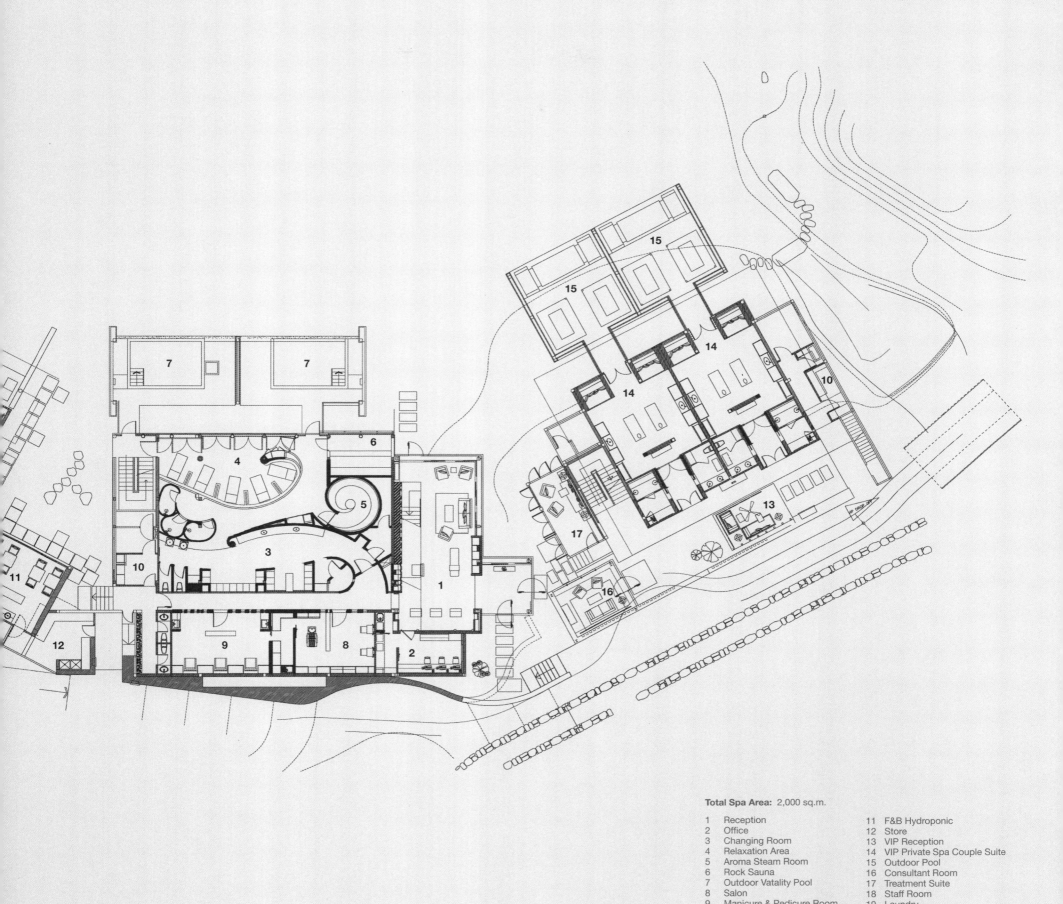

Total Spa Area: 2,000 sq.m.

1	Reception	11	F&B Hydroponic
2	Office	12	Store
3	Changing Room	13	VIP Reception
4	Relaxation Area	14	VIP Private Spa Couple Suite
5	Aroma Steam Room	15	Outdoor Pool
6	Rock Sauna	16	Consultant Room
7	Outdoor Vatality Pool	17	Treatment Suite
8	Salon	18	Staff Room
9	Manicure & Pedicure Room	19	Laundry
10	Linen & Prep Room	20	Pond

The Barai

Location
Hyatt Regency Hua Hin
91 Hua Hin - Khao Takiap Road, Hua Hin,
Prachuap Khiri Khan 77110
T: +66 (0) 3252 1234 F: +66 (0) 3252 1233
E: thebarai.hrhuahin@hyatt.com
www.thebarai.com

Design Firms
Architect and Interior Designer: Bunnag Architects
                                          International
                                          Consultants
Landscape Architect: Bensley Design Studios
Lighting Designer: Vision Design Studio

Sun light and shadow are key element inspiring the design concept of The Barai. The architectural design aims to create the tranquility and delightful atmosphere that is the spirituality of the East. Visitors would intuitively feel an inner peace along the journey through its enclosed open space and mysterious hallways as the movements of sun and shadow play with the varieties of form and space.

The designer deliberately characterizes the spa as mysterious and calm with the use of strong light and shadow. The colourful skylight glass is employed to invite the direct and amplified sunlight to penetrate the narrow space deep into the dark room to create multiple shades of colour reflected on wall, floor, and ceiling surfaces.

Besides this bounding edge, the swimming pool is designed to be surrounded by eight-meter high walls serving as wind protectors so the water surface remains calm, creating reflections of realistic clarity.

The design of this walled compound is inspired by Khmer and Thai cultural heritage, with a contemporary touch. The architectural technique employs a modulation of room types featuring massive enclosing walls, symmetrical spaces, uneven room heights and variations of space. The use of architectural connectors, a Western designing language, is applied to integrate all the rooms. Some connectors have no roof while in another place an open courtyard appears, encompassing the existing big trees as part of its architectural aesthetic.

Overall, the design achieves a sense of privacy and the opportunity for visitors to be with the nature in an artistic relationship. The Barai is not only designed for commercial purposes but also for social culture, benefitting both individual and community.

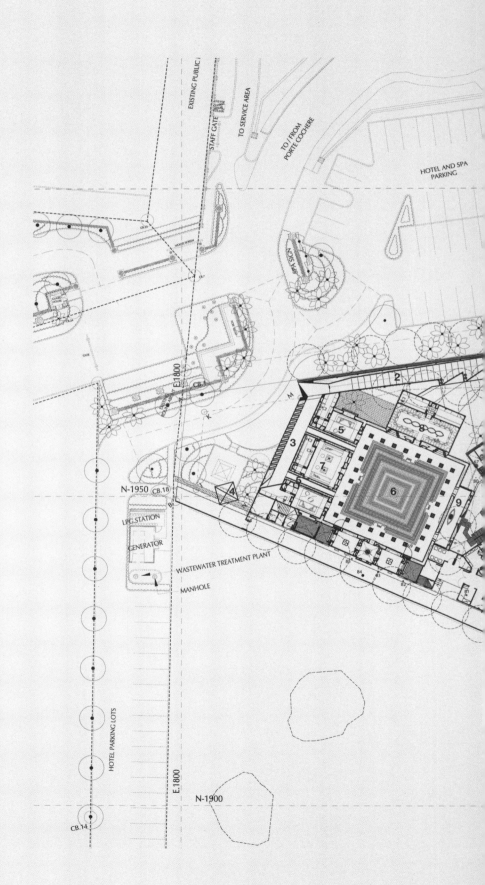

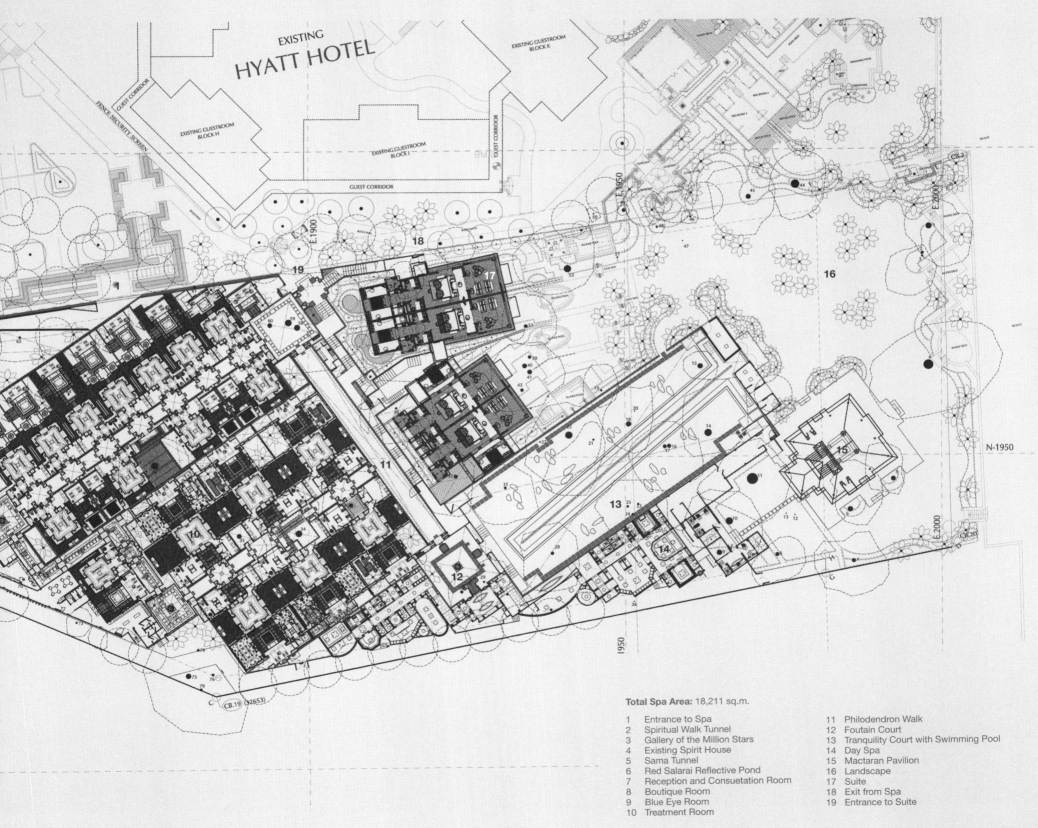

Total Spa Area: 18,211 sq.m.

1 Entrance to Spa
2 Spiritual Walk Tunnel
3 Gallery of the Million Stars
4 Existing Spirit House
5 Sama Tunnel
6 Red Salarai Reflective Pond
7 Reception and Consuetation Room
8 Boutique Room
9 Blue Eye Room
10 Treatment Room

11 Philodendron Walk
12 Foutain Court
13 Tranquility Court with Swimming Pool
14 Day Spa
15 Mactaran Pavilion
16 Landscape
17 Suite
18 Exit from Spa
19 Entrance to Suite

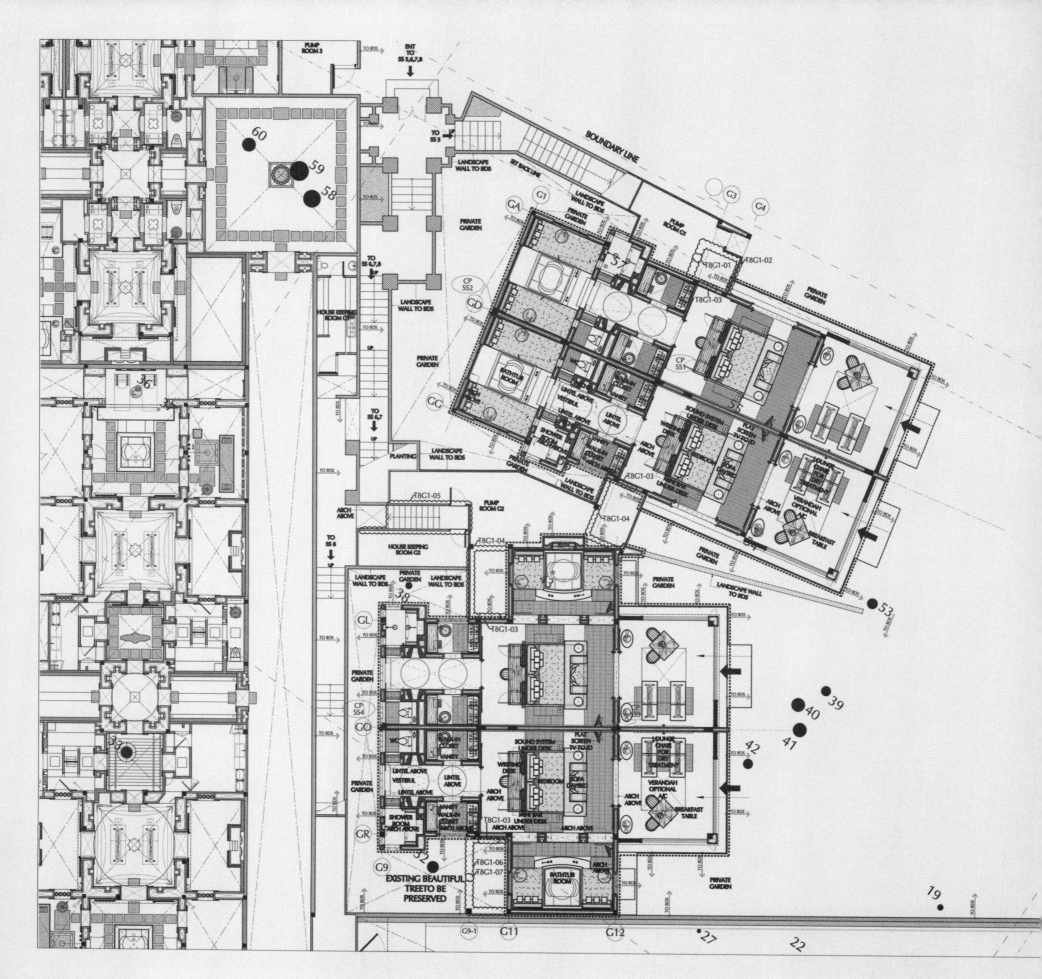

Spa Suite

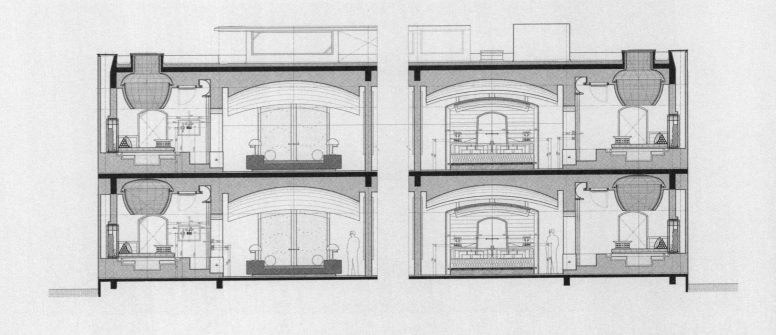

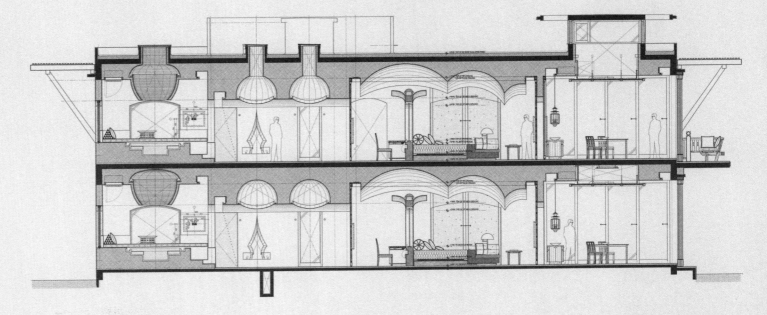

Spa Suite Section

SALA Spa

Location
Sala Phuket Resort and Spa
333 Moo 3 Mai Khao Beach Thalang,
Phuket 83110
T: +66 (0) 7633 8888   F: +66 (0) 7633 8889
E: info@salaphuket.com
www.salaresorts.com

Design Firms
Architect and Interior Designer: Department of
                                   Architecture in
                                   collaboration with
                                   Pakorn Mahapant

Soft Furnishing and Interior
Accessories Designer: Gasinee Chieu Jirathitwanitkun
Lighting Designer: Accent Studio

As a continuation of the overall design idea for the SALA Phuket Resort, which ties into memories and traces of old town Phuket, SALA Spa integrates but reinterprets the spatial sequences of Phuket's old shop houses — a linear sequence of spaces alternating between the different levels of shades and shadows, between darkness and light, and between the enclosed internal spaces and the outdoor courtyards open to the sky, wind, and rain. The pathway slowly reveals layers of unexpected experiences.

Water plays an important role within the spa complex. The passage takes the visitor through varying forms of water in each space; a still and silent reflecting pond, the soothing sound of a gurgling stream flowing through small rocks, drops of rain, the rippling of light reflected off the water onto the surface of the wall, and even the atmosphere of the air itself, filled with mist.

Concrete walls, which define the passageway, feel surprisingly natural, formed by timber formworks that reveal the texture and pattern of the wood. This gives a subtle materiality that lies somewhere between wood and concrete. The wall texture reveals itself differently with the varying angles of light in different locations and at different times of day.

On entering the spa complex, it is almost as if one is moving into a different world down this long and narrow path. Arriving at the lobby, the space opens up. There is light, trees, and the sound of water flowing slowly through rugged pebbles. From here to the spa treatment rooms, one disappears again into the passage of mist. After the treatment, and the time spent lingering in the spa, one emerges to discovers that the quality of light has changed. The perception of the same space is not the same.

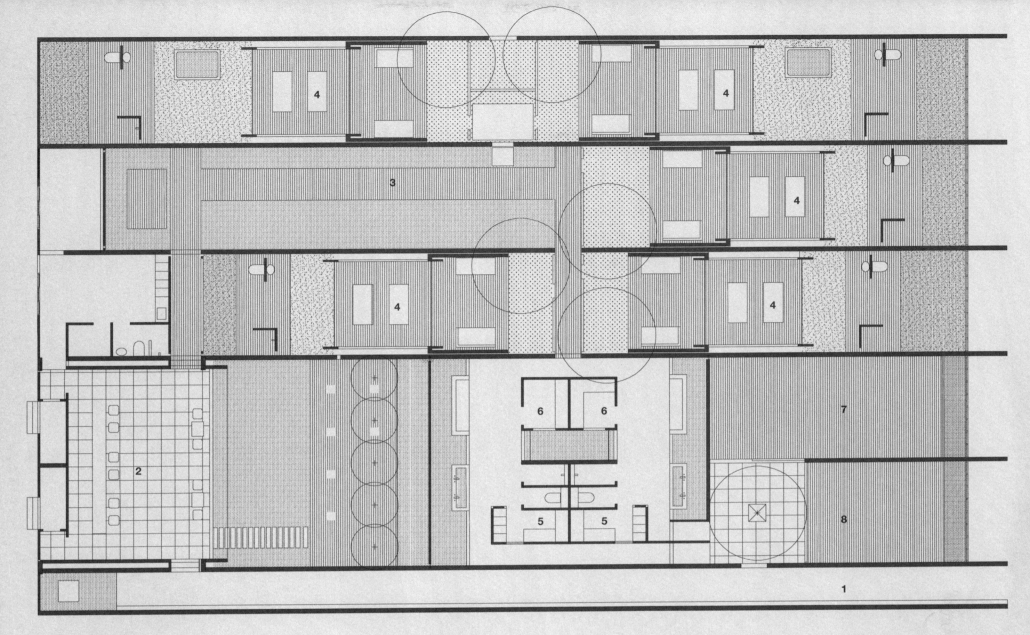

Total Spa Area: 1,200 sq.m.

1	Entrance	5	Locker
2	Lobby	6	Steam Room
3	Catwalk Hall	7	Fitness
4	Spa Room	8	Salon

Section

Spa Naka

Location
The Naka Island, A Luxury Collection Resort & Spa
32 Moo 5 Paklok, Thalang, Phuket 83110
T: +66 (0) 7637 1400 F: +66 (0) 7632 7338-9
E: naka.reservations@luxurycollection.com
www.luxurycollection.com/naka

Design Firms
Architect: Habita Architects
Landscape Architect: P Landscape

Nestled among stunning beaches and lush coconut groves, with never-ending views of emerald-green Phang Nga Bay and the idyllic landscape of the Phuket coastline, The Naka Island is an exclusive boutique resort on Naka Yai Island, located just off the Phuket East coast. While just 25 minutes from Phuket International Airport, The Naka Island is accessible only by private speedboat, making it a uniquely private and intimate retreat. The resort offers sixty-seven villas that are magnificently appointed with private plunge pools and tropical gardens, along with an incomparable spa facility, access to countless outdoor recreation opportunities, and uncompromising service.

At a spacious 770 square meters, Spa Naka by Luxury Collection is one of the largest spas in the region. Spa Naka is an exclusive five-star sanctuary that focuses on the key principles of health, beauty and inner peace. With The Naka Island's convenient yet completely private location, just five minutes from the East coast of Phuket, Spa Naka is the perfect retreat for guests of the resort as well as visitors to Phuket looking to get away for a luxury experience.

Spa Naka's design is inspired by the rich indigenous culture, with a layout resembling a quiet Thai village. The buildings and walkways are constructed of natural stone along serene paths that meander through lush foliage and over water features, including a large Watsu pool used for aquatic therapy. The spa offers twelve private treatment suites, including four couples suites that feature outdoor rainforest showers, a private plunge pool, a steam room and steam bed. Ice room and Kanieep pool, a hot and cold pool that is particularly designed for foot massages, are available.

The Relaxation Garden offers quiet seating and an opportunity to enjoy Spa Naka's organic native plantation, where, featuring a variety of herbs for special healing treatments, the Spa Naka therapists will hand select fresh herbs to incorporate with treatments based on personalized service. Organic herbal tea is then served following treatments.

Spa Naka offers a collection of luxurious treatments that pamper from head to toe, especially the signature Watsu treatment, a 60-minute-aquatic therapy that combines shiatsu pressure points, gentle stretches and the absence of gravity to help alleviate those aches and pains to keep guest on track for the challenges of day to day life. The unique Watsu experience is suitable for all ages, genders and swimming abilities. Other unique discovering rituals are accessible and ready to be experienced.

Total Spa Area: 770 sq.m.

1 Spa Reception
   1.1   Reception
   1.2   Waiting 1
   1.3   Waiting 2
   1.4   Spa Product Storage
2 Changing Room
   2.1   Steam
   2.2   Ice Room
   2.3   Sauna
   2.4   Kneipp
   2.5   Bathroom
   2.6   Dressing
   2.7   Relaxation Area
3 Treatment Room
4 Watsu Pool
5 Relaxing Sala
6 Treatment Room
7 Beach Pool Villa Suite
8 Existing Big Tree
9 Bamboo Tunnel Entrance

Coqoon Spa

Location
Indigo Pearl Phuket
Nai Yang Beach and National Park, Phuket 83110
T: +66 (0) 7632 7006, +66 (0) 7632 7015
F: +66 (0) 7632 7338-9
E: info@indigo-pearl.com
www.indigo-pearl.com/Coqoon-Spa

Design Firms
Interior Designer, Architect
and Landscape Architect : Bensley Design Studios
Lighting Designer : Withlight

As a specialist in styling exotic Asian luxury resorts, Bangkok-based architect and landscape designer Bill Bensley is not for the feint-hearted. "I like things quirky, a bit strange and the odder the better," he says and his off-beat mantra has been eloquently demonstrated over the years at some of Asia's most praised hotels, including the acclaimed Indigo Pearl in Phuket.

Widely applauded as a design masterpiece, scooping up 'Asia's Leading Design Hotel' award at the World Travel Awards in 2012, and praised by Conde Nast Traveler, HAPA and others, Bensley has taken Indigo Pearl's 'raw-industrial meets sensual-chic' styling to another level with the unveiling the cutting-edge Coqoon Spa.

Modeled on his boyhood vision of a 'Garden of Eden', Coqoon Spa boasts six individual treatment rooms and a lavish spa suite complete with private pool, Vichy rain shower, steam room and sauna. Its pièce de résistance is 'The Nest' – the spa's signature treatment suite suspended midair in the branches of what is thought to be Phuket's oldest banyan tree.

Accessed by a 'flying bridge', the 'Nest' allows guests to get up close and personal to nature. The internationally renowned designer is the first to admit that having designed many spas, none are like this one. "I consider myself a designer of environments. There are no boundaries. As a landscape architect I've been taught to protect the earth and environment, and as an architect I want to do the same," he continues. 'The Nest' is undoubtedly unique. Hand-woven and hovering 10-metres up in the branches, he says it satisfied his dream of doing something Bensley and quirky, something distinct from any other place in Thailand. It also aligns with his philosophy, summed up by the Indonesian phrase lebih gila, lebih biak, which means 'the more odd, the better'.

As a landscape architect, he is also extremely eco-sensitive, with much of his work based on eco-nature. "I believe that respecting Mother Nature and the natural sites we are given is my calling in life as a landscape architect. Years from now the banyan tree will grow around our cocoon and embrace it in its own way. The interplay of this gorgeous tree is paramount. The chic aspect is secondary. I love resorts where you can still find a few surprises, new details on your fourth or fifth day."

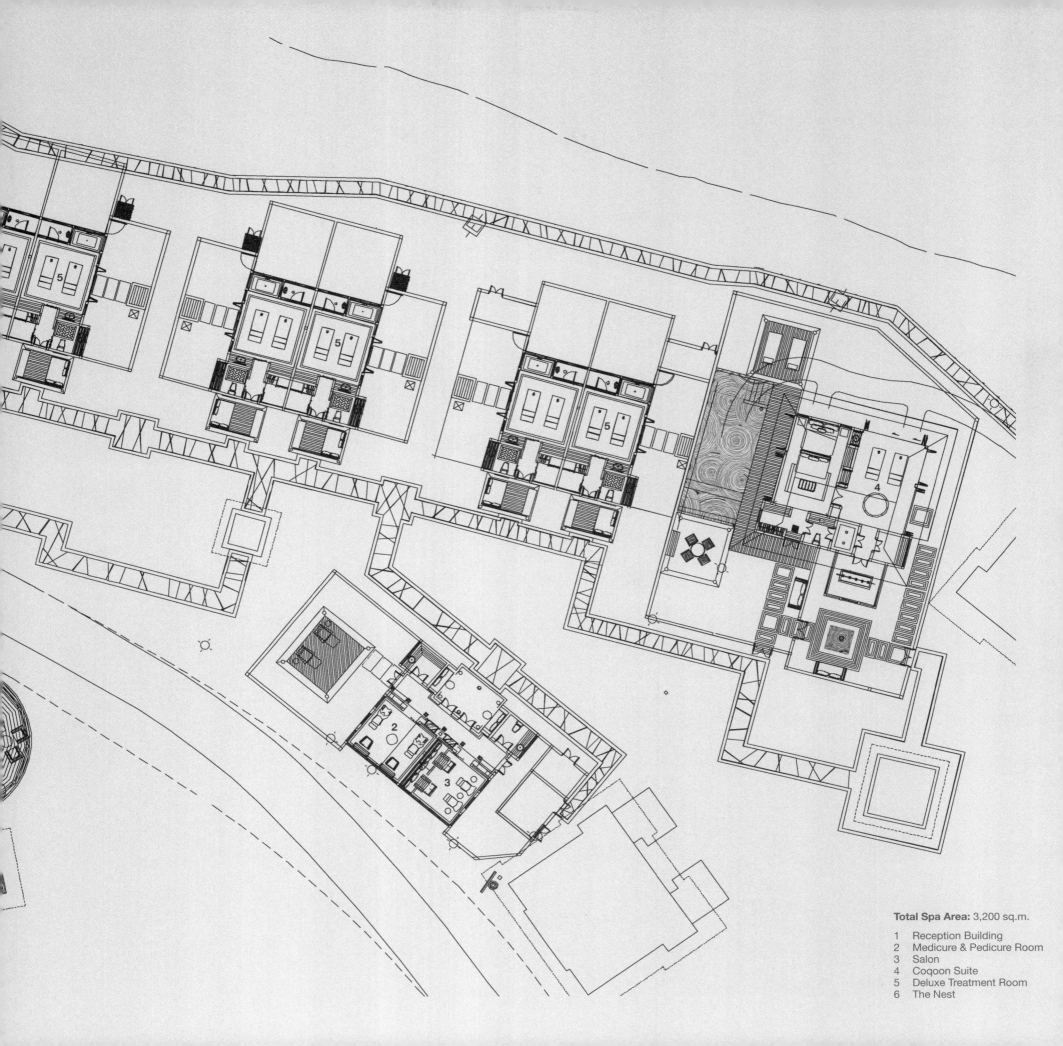

Total Spa Area: 3,200 sq.m.

1   Reception Building
2   Medicure & Pedicure Room
3   Salon
4   Coqoon Suite
5   Deluxe Treatment Room
6   The Nest

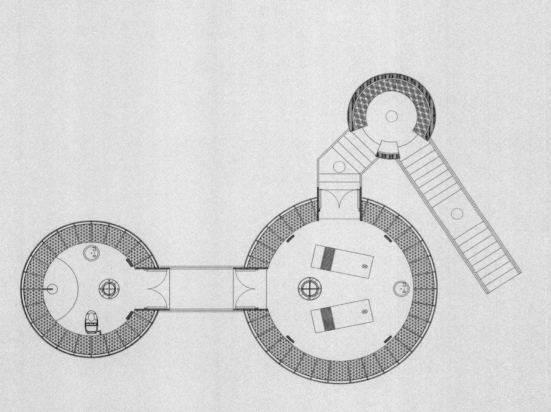

The Nest

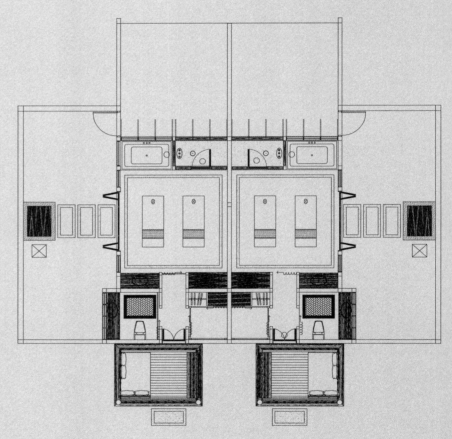

Deluxe Treatment Room

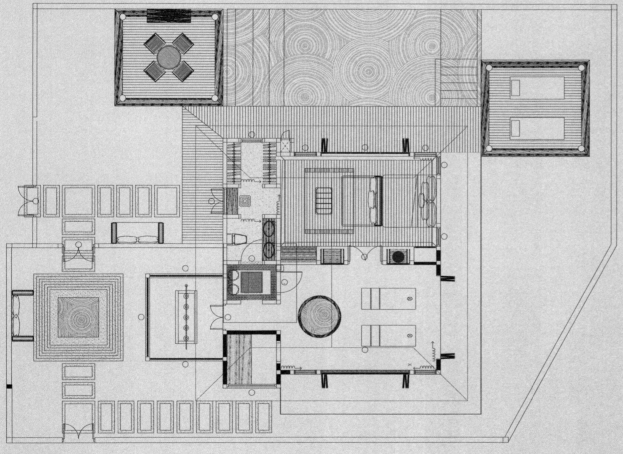

Coqoon Suite

The Forest Spa

Location
Tamarind Spring
205/7 Mu 4, Thong Takian, Koh Samui,
Suratthani 84310
T: +66 (0) 7742 4221, +66 (0) 7723 0571
F: +66 (0) 7742 4311
E: spa@tamarindsprings.com
www.tamarindsprings.com

Design Firm
Architect and Interior Designer: Detlef Dirksen

Tamarind Springs Forest Spa is Koh Samui's first day spa, an award-winning and eco-certified oasis featuring giant granite boulders and an exquisite tropical hillside grove, where guests slow down to reconnect with the nurturing rhythm of nature and experience the tranquility of this magical place.

One of Asia's most acclaimed day spa destinations, Tamarind Springs specializes in an exclusive indulgence of all the senses: stepping through the gently landscaped forest property has been described as a 'walking mediation experience', which, after a welcoming introduction, is followed by a extended visit to the spa's fragrant herbal steam caves and cool rock pools.

With the mind now calm guests move on to enjoy one of several superb massage options in private open-plan massage pavilions. The professional therapists have been internationally trained, in traditional Eastern as well as Western techniques, to deliver effective treatments having the dual benefit of imparting a state of relaxation and of enabling healing in a holistic way.

The beginnings of Tamarind Springs go back to the late 80s. It was then that the original concept developed out of a rich grassroots herbal heritage and a genuine tradition of massage handed down through generations. In addition, service is an intrinsic Thai attribute and is delivered at Tamarind Springs with genuine unpretentiousness and natural friendliness. And to make it perfect, the experience is enhanced by Tamarind Springs' unique environment, where sights, smells and sounds combine to heighten the whole experience of sabai (feeling good, relaxed).

This unique landscape coupled with the ancient tradition of Thai massage provided the inspiration for Tamarind Springs Forest Spa. Turning inspiration into reality and integrating structures and features led to two herbal steam caves, three rock plunge pools, open private massage pavilions scattered across the hillside, a yoga sala, and a large reception and café building.

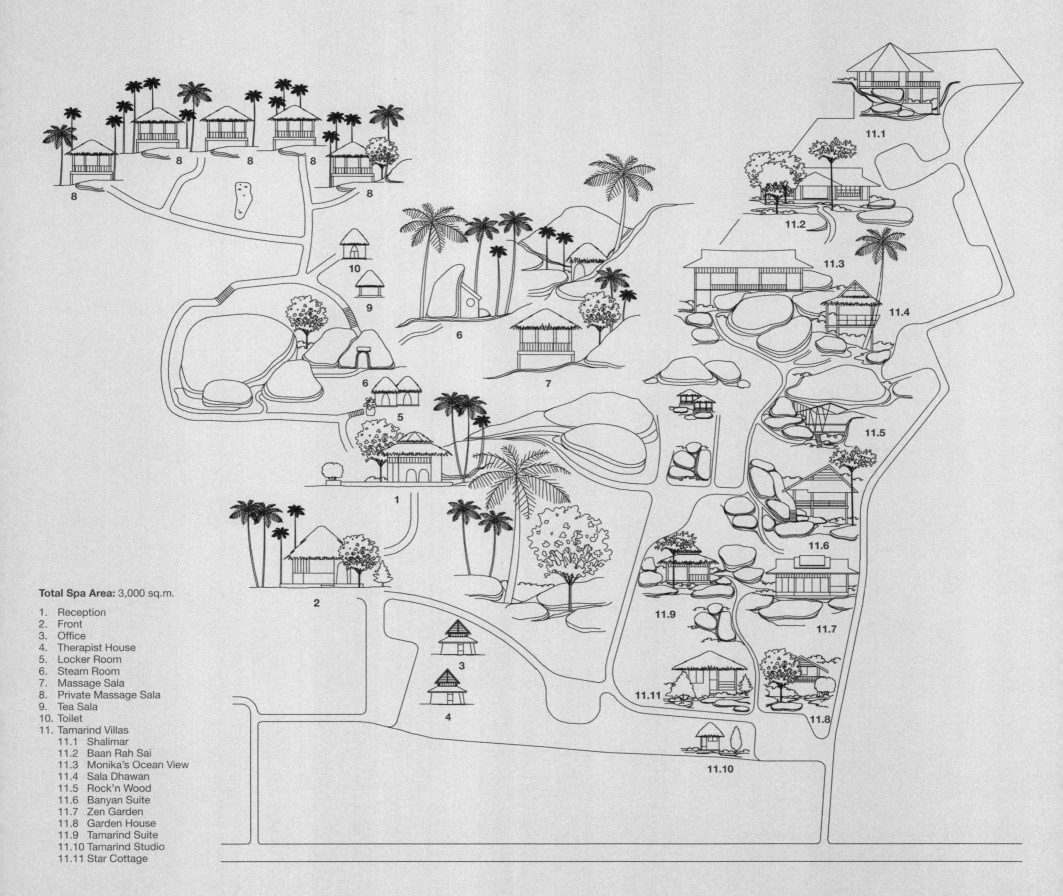

Total Spa Area: 3,000 sq.m.

1. Reception
2. Front
3. Office
4. Therapist House
5. Locker Room
6. Steam Room
7. Massage Sala
8. Private Massage Sala
9. Tea Sala
10. Toilet
11. Tamarind Villas
    11.1 Shalimar
    11.2 Baan Rah Sai
    11.3 Monika's Ocean View
    11.4 Sala Dhawan
    11.5 Rock'n Wood
    11.6 Banyan Suite
    11.7 Zen Garden
    11.8 Garden House
    11.9 Tamarind Suite
    11.10 Tamarind Studio
    11.11 Star Cottage

Top Exotic Spa in Thailand

Photography Credit
t = Top, b = Bottom l = Left, r = Right, c = Center

Christopher Cypert	ESPA
Kan Sivapuchpong	So Spa
Markus Gotz	pp. 148t, 149
Mathar Bunnag	pp. 142-145, 146r
Prakit Phananuratana	Lotus Series (pp. 10-11, 24-25, 36-37, 48-49, 62-63, 74-75, 86-87, 98-99, 110-111, 122-123, 136-137, 152-153, 168-169, 184-185, 198-199, 212-213, 226-227)
Srirath Somsawat	Kempinski The Spa (pp. 80-86, 88-89), Thann Sanctuary (pp. 108-111), The Spa at The Chedi (pp.116-121, 124-125) and Coqoon Spa

All other photos courtesy of listed Architects , Spas and Hotels.

Acknowledgements

On behalf of Li-Zenn Publishing Limited, we would like to take this opportunity to thank all who co-operated in the successful production of this publication. First of all, thanks to the 17 Spas and Hotels who participated in the project and generously supported with images and information. And a special thank you to Coqoon Spa which provide accommodation and helped facilitate the photo shoot. Also, we extend our gratitude to all the architects and interior designers from Aesthetics Architects, Architect 49 Phuket, August Design consultant, Bensley Design Studio, Bunnag Architects International Consultants, Department of Architecture, Interior Architect 49, PIA Interior, P49 Deesign and Habita Architects, who supported the project with great images and information. We are further grateful to Matha Bunnag, who offered great advice, to Ander Jacker, the President of Thai Spa Associates, who kindly wrote the Foreword, and to Mayitka Tangkaewfa who provided spa contacts and gathered information.
Lastly, thanks to Adam Smith for his creative ideas and delicate touch in the making of this publication.